PROBLEM SOLVING FOR ENGINEERS AND SCIENTISTS

A CREATIVE APPROACH

PROBLEM SOLVING FOR ENGINEERS AND SCIENTISTS

A CREATIVE APPROACH

Raymond Friedman

CHAPMAN & HALL

I(T)P An International Thomson Publishing Company

New York • Albany • Bonn • Boston • Cincinnati • Detroit • London • Madrid • Melbourne •
Mexico City • Pacific Grove • Paris • San Francisco • Singapore • Tokyo • Toronto • Washington

This edition published by Chapman & Hall, New York, NY

Printed in the United States of America

For more information contact:

Chapman & Hall
115 Fifth Avenue
New York, NY 10003

Chapman & Hall
2-6 Boundary Row
London SE1 8HN
England

Thomas Nelson Australia
102 Dodds Street
South Melbourne, 3205
Victoria, Australia

Chapman & Hall GmbH
Postfach 100 263
D-69442 Weinheim
Germany

Nelson Canada
1120 Birchmount Road
Scarborough, Ontario
Canada M1K 5G4

International Thomson Publishing Asia
221 Henderson Road #05-10
Henderson Building
Singapore 0315

International Thomson Editores
Campos Eliseos 385, Piso 7
Col. Polanco
11560 Mexico D.F.
Mexico

International Thomson Publishing - Japan
Hirakawacho-cho Kyowa Building, 3F
1-2-1 Hirakawacho-cho
Chiyoda-ku, 102 Tokyo
Japan

3 4 5 6 7 8 9 XXX 01 00 99 98 97 96

Library of Congress Cataloging-in-Publication Data

Friedman, Raymond.
        Problem solving for engineers and scientists : a creative approach / Raymond Friedman.
            P.   Cm.
        Includes index.
        ISBN 0-442-00478-8
        1. Engineering mathematics. 2. Science--Mathematics. 3. Problem solving. I. Title.
        TA330.F76   1991   90-25154
        620'.0042--dc20   CIP

Visit Chapman & Hall on the Internet http://www.chaphall.com/chaphall.html

To order this or any other Chapman & Hall book, please contact **International Thomson Publishing, 7625 Empire Drive, Florence, KY 41042.** Phone (606) 525-6600 or 1-800-842-3636. Fax: (606) 525-7778. E-mail: order@chaphall.com.

For a complete listing of Chapman & Hall titles, send your request to **Chapman & Hall, Dept. BC, 115 Fifth Avenue, New York, NY 10003.**

CONTENTS

Foreword   ix

Chapter 1   WATER, WATER EVERYWHERE   1
      1.1   Of Time and the River: Speedboats   1
      1.2   Of Time and the River: Sailing   1
      1.3   Floating on the Pond   2
      1.4   The Bubble in the Water Tank   3
      1.5   The Bottom of the Lake   3
      1.6   When Will the Puddle Evaporate?   3

Chapter 2   SOME "ELEMENTARY" PROBLEMS   24
      2.1   When Do I Add the Cream to my Coffee?   24
      2.2   The Hot-Air Balloon   25
      2.3   How Far Can an Airplane Fly?   26

Chapter 3   WHAT IS THE CHANCE THAT . . . ?   37
      3.1   Introduction   37
      3.2   Division By 137   38
      3.3   The Turbine Seals   39
      3.4   The Keys to the Taxicabs   40

Chapter 4 UNSTABLE BEHAVIOR   45
    4.1 Introduction   45
    4.2 The Solid-Propellant Rocket   48
    4.3 Boiling Water With a Hot Wire   49
    4.4 A Catalytic Particle   49

Chapter 5 FINDING THE OPTIMUM   62
    5.1 Introduction   62
    5.2 Draining the Tank   62
    5.3 Insulating the Electric Cable   63
    5.4 Minimizing Automobile Collisions   64
    5.5 Hitting a Golf Ball   64

Chapter 6 SOME SCIENTIFIC CURIOSITIES   83
    6.1 Surface-to-Volume Ratio   83
    6.2 The Leaning Tower of Pisa   83
    6.3 Absolute Temperature   84
    6.4 A Matter of Gravity   84
    6.5 Letting Air Into an Empty Tank   86
    6.6 The Two Capacitors   86

Chapter 7 VIOLATING THE LAWS OF THERMODYNAMICS   101
    7.1 Introduction   101
    7.2 Underwater Rotating Device   102
    7.3 Electrolysis of Water   104
    7.4 A Submarine That Needs No Fuel   105
    7.5 Steam and Ethylene Glycol   107
    7.6 An Odd Case of Radiative Exchange   109

Chapter 8 SOME PROPAGATION PROCESSES   123
    8.1 A Weak Shock Wave   123
    8.2 Melting Through Ice   123
    8.3 Combustion Wave in a Thermite Mixture   124
    8.4 Flame Propagation in a Combustible Gas   124

Chapter 9   A LECTURE ON DIMENSIONLESS GROUPS   142

9.1   Mach Number   144

9.2   Froude Number   145

9.3   Reynolds Number   146

9.4   Grashof Number   147

9.5   Euler Number and Bernoulli Equation   148

9.6   Barometric Formula   149

9.7   Weber Number   149

9.8   Other Dimensionless Numbers   149

Index   155

FOREWORD

Let's assume that you, the reader, have been educated in the basics of science and perhaps some branch of engineering. You have access to textbooks and handbooks, and you are comfortable with a computer. One day, you find yourself faced with a technical problem.

If this problem happens to be essentially identical to a previous problem, which you were trained to solve, clearly you will have no difficulty. Unfortunately, however, the variety of technical problems that you might encounter is enormous. The chance is small that a given problem will be an exact replica of a familiar problem.

In such a case, it is possible that you will reach out for the first relevant quantitative relationship that occurs to you. You then might substitute numbers into a formula and obtain an answer, perhaps with the aid of a computer or hand calculator.

As the many examples in this book will demonstrate, your answer is quite likely to be wrong because you have overlooked some important aspect of the problem. You can easily convince yourself of this fact by tackling some of the problems in this book.

If this assertion is true, what then? You must then admit that you are not very effective in solving unfamiliar problems. It is hoped that at this point you will feel challenged to improve your problem-solving ability. The goal of this book is to help you to do so. Rather than lecture to you on abstract principles of problem solving, I wish to give you a chance to learn by experience.

The book consists of about 35 problems, each dealing with some simplified physical situation. After you have read the statement of a problem, you will be given a chance to solve it yourself, or at least to formulate a general approach to the solution. Then you will study a dialogue among a professor and several students. The students propose solutions, and the professor provides criticisms and guidance toward the correct solution. After several false starts, ultimately a valid solution emerges.

I have accumulated these problems during a long career in multidisciplinary engineering research. They have been selected from a multitude of other possible problems because each meets most of the following criteria:

- A fairly simple statement of the problem is possible.
- The problem involves a physical situation rather than a mathematical puzzle or paradox.
- A unique solution can be deduced without complex mathematics.
- It is very easy to come up with a wrong solution.
- The problem is not easily found in common textbooks.

The art of solving these problems is best appreciated if you first try your hand and then study the correct solution as presented in the dialogue. Finally, you should ask yourself where you went wrong (if you did). In some cases, of course, you may not have known the crucial bit of factual information needed. While working your way through this book, you will pick up some bits of information that may be useful in the future.

Although the book tries to avoid lecturing the reader on general principles of problem solving, because this makes for dull reading, nevertheless it may be of interest to some readers to see a list of helpful guidelines for solving technical problems.

1. Question the statement of the problem. Is it clear? Does it have any element of ambiguity? Is some crucial information missing?

2. Don't assume that the problem is insoluble simply because you are unable to solve it in 15 minutes. That may be true of a problem in a school examination with a time limit, but the solutions of real-world problems often justify and require study for hours or days. If your

first approach doesn't work, be mentally prepared to try other methods.

3. If you are not too knowledgable in the subject matter of the problem, it may be very helpful first to do some background reading in an appropriate textbook or encyclopedia in order to become aware of some crucial, relevant fact.

4. One way to solve a problem is by a series of successive refinements. First, outline a train of logic that yields a qualitative description of the solution. Next, make the crudest possible quantitative analysis to reveal the order of magnitude of the answer. Finally, develop a more precise solution. However, remember that there are many real-world problems for which the input data are never precisely known; for example, the trajectory of a projectile as influenced by the wind. In such cases a very precise solution is not needed.

5. Once you obtain a solution, test it in various ways. First, ask if it is consistent with basic laws such as conservation of mass, energy, and momentum and the second law of thermodynamics.

6. Another test, of course, is whether the solution is consistent with "common sense." However, as examples in this book will show, the correct answer is sometimes at variance with the intuitive, or "commonsense," answer.

7. If the solution is a mathematical formula, it may be tested in various ways. Substitute extreme values for the independent variables and see if the dependent variable takes on reasonable values. (In other words, does the solution satisfy physical boundary conditions?) Check to see if any term in the denominator can go to zero with certain inputs, causing the independent variable to go to infinity, when this result is not physically reasonable. (In other words, can the solution "blow up"?) It is often helpful to plot the solution on graph paper.

8. Be aware that there is sometimes more than one correct way to solve a problem. If you are aware of two ways, choose one. Then try the other way, as a check.

9. Often, in solving a problem, you will choose to neglect certain factors that you hope are unimportant. Once you obtain a solution, it is desirable to calculate the sensitivity of the result to such neglected factors.

10. It is often very helpful to discuss the problem with another qualified person (a consultant), if available, both before and after you obtain your solution.

As you work through this book, you will see that the problems are usually mechanical, thermal, chemical, or electrical in nature. In no case is mathematics beyond elementary calculus needed. Several sections can be dealt with more comfortably if you have some prior exposure to fluid mechanics or thermodynamics. No mention is made of relativity or quantum mechanics. The underlying goal of each problem is to predict behavior in the real world as governed by fundamental scientific principles.

CHAPTER 1

WATER, WATER EVERYWHERE

1.1 OF TIME AND THE RIVER: SPEEDBOATS

A speedboat race is about to start from a small island in the middle of a wide river, which flows from north to south. Some of the racers are told to go south, to go around a buoy that is 1 mile downstream of the island, and then to return to the island. The remainder of the racers are told to go east (crosscurrent), to go around another buoy that is also 1 mile away, and then to return to the island. Thus, all racers must cover a distance of 2 miles. (See Figure 1.1A on page 2.)

The water is flowing downstream at a constant velocity. All speedboats are capable of a speed greater than this flow.

A question is raised about the fairness of this procedure. Does either group of racers have an advantage?

READER: Before turning to page 5, provide your solution to the problem.

1.2 OF TIME AND THE RIVER: SAILING

A wide, straight river flows from north to south. The current is constant at 5 miles per hour. I have a small sailboat, and I wish to sail south, to a town 30

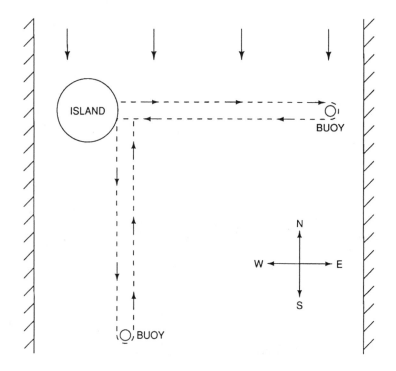

Figure 1.1A   The boat race.

miles downstream. I wish to consume no more time than necessary during this one-way trip. I can either start this morning or tomorrow morning.

The weather bureau informs me that the wind is blowing due south at 5 miles per hour and will continue at this rate all day today. Tomorrow, there will be no wind whatsoever.

Can I accomplish the trip in less time if I start now? Or tomorrow?

READER: Solve the problem. Then turn to page 7 for the solution.

1.3 FLOATING ON THE POND

I am in a canoe, floating on a small pond. In the canoe with me is a large rock. I drop the rock overboard, and it sinks to the bottom of the pond. This is all the information that will be supplied.

Does the water level in the pond rise, fall, or stay the same?

READER: Provide your answer, along with your reasoning, before turning to page 8 for the solution.

1.4 THE BUBBLE IN THE WATER TANK

A heavy-walled, vertical, cylindrical steel tank, 10.3 meters high, is filled with water and pressurized until the pressure at the top is 2 atmospheres absolute, whereas the pressure at the bottom is 3 atmospheres absolute. (This 1-atmosphere difference is caused by the hydrostatic head of a column of water 10.3 meters high.) The line from the pump to the tank is then closed off; i.e., the tank is sealed.

There is a rupture diaphragm in the bottom of the tank that will rupture at 2.8 atmospheres differential pressure (i.e., 3.8 atmospheres absolute pressure inside). There is an air bubble in the water, adhering to the bottom of the tank.

That is the initial situation. Then the air bubble detaches itself from the bottom of the tank and rises to the top. (See Figure 1.4A on page 4.)

Does this movement cause the diaphragm to rupture?

READER: What do you think would occur? Why? Then turn to page 9 for the solution.

1.5 THE BOTTOM OF THE LAKE

A certain lake in Canada is 100 meters deep. The lake contains pure water. There are no nearby geothermal heat sources.

On next June 1, what will be the temperature of the water at the bottom of the lake?

READER: Do you care to take a stab at this one? Then turn to page 12 for the solution.

1.6 WHEN WILL THE PUDDLE EVAPORATE?

I spill 100 grams of water on my large dining-room table. The water forms a circular puddle 0.3 meters in diameter.

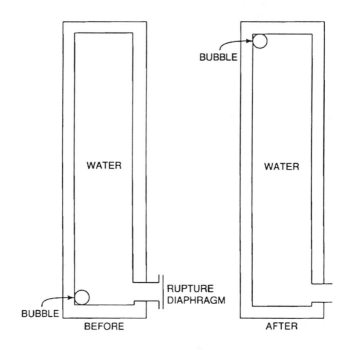

Figure 1.4A   The bubble in the water tank.

Estimate how long it takes to evaporate completely. What additional information do you need, if any, to solve this problem?

READER: List the information you think is needed, in addition to what may be found in engineering handbooks, before turning to page 14.

SOLUTIONS TO CHAPTER 1
WATER, WATER EVERYWHERE

1.1 OF TIME AND THE RIVER: SPEEDBOATS

Professor: Who has a solution to the problem?

First Student: It seems fairly obvious to me. The racers heading for the buoy that is due east of the starting point, and therefore moving at right angles to the current, are going to be carried somewhat downstream before they reach the buoy, unless they compensate by heading somewhat upstream as they go. Therefore they will not make as good time as they would have if moving due east on still water. On the return trip to the island, going west, they will again have the same problem.

On the other hand, the racers heading for the south buoy, going downstream, are helped by the current on the outbound leg and hindered by the current on the return leg. These effects should cancel one another, and the net result would be the same as if they were moving on still water. Therefore it seems clear that the racers heading for the south buoy have a clear advantage.

Professor: If you worked for me, and I reduced your salary by 30 percent today and then increased it by 30 percent tomorrow, would you then be making the same salary as you started with? In fact, you would soon discover that your salary was 9 percent lower. The present problem requires a mathematical formulation. I'll give you 5 minutes to work it out.

Second Student: I have a solution. Let U be the speed of a motorboat on still water and let V be the speed of the current, each in miles per hour. Each leg is a distance of 1 mile. First, consider the southbound racers. They will require $1/(U + V)$ hours to reach the buoy and $1/(U - V)$ hours to return. The total time is the sum of these quantities:

$$1/(U + V) + 1/(U - V) = \frac{2U}{U^2 - V^2} = \frac{2}{U[1 - (V^2/U^2)]}$$

5

Thus, the time is greater than $2/U$ as long as V/U is greater than zero (finite current), and the time becomes infinite when V/U becomes as large as one. At such a condition, the motorboat can make no progress upstream.

Now we must calculate the time required for the eastbound racers. Here we use a vector diagram (Figure 1.1B). A vector of magnitude U, representing a boat, points somewhat north of east, to compensate for the tendency of the current to cause the boat to drift south as it tries to proceed east. A separate vector of magnitude V, representing the current, points due south. The resulting velocity of the boat is the sum of these two vectors, which is one leg of a right triangle. The hypotenuse is U and the other leg is V. Therefore the resulting velocity, by the Pythagorean theorem, is $\sqrt{(U^2 - V^2)}$.

As for the westward return trip, the situation is exactly the same, so the velocity is the same as for the eastward trip. The time required for the round trip, then, is $2/\sqrt{(U^2 - V^2)}$. This may also be expressed as $2/\{U\sqrt{[1 - (V^2/U^2)]}\}$.

This formula tells us that the east-west round-trip time is $2/U$ when V is zero, as we would expect, and also that the round-trip time is infinite if V is as large as U.

Finally, to answer the original question, we take the ratio of the east-west round-trip time to the south-north round-trip time, and we obtain $\sqrt{[1 - (V^2/U^2)]}$. For any value of V/U between zero and 1, which is the entire range of interest, the ratio is seen to have a value less than 1. This result means that the motorboats going east-west have a clear advantage over those going north-south, which is the opposite of the conclusion reached by the first student.

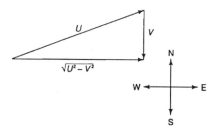

Figure 1.1B   Vector diagram for boat going eastward.

Professor: If you did that entire analysis in 5 minutes, maybe you should be the teacher.

Second Student: Actually, I was reading a book about the theory of relativity last night. The Michelson-Morley experiment involving the velocity of light through the so-called ether was described, and essentially the same equations were involved.

1.2 OF TIME AND THE RIVER: SAILING

Professor: I know you all suspect that there must be a trick to this problem. There is. So think carefully, and then give me your answer.

First Student: This is the way I see it. If I were to make the journey tomorrow, when there is no wind, I obviously would drift with the current at 5 miles per hour and accomplish the 30-mile journey in exactly 6 hours. That much is clear.

If I were to start this morning, both the water and the wind would be moving south at 5 miles per hour, relative to the shore, so the velocity of the wind relative to the water would be zero. If my sailboat is drifting with the water at 5 miles per hour initially, and I erect my sail, I would find that the wind has no velocity relative to my sail, so it would not benefit me at all.

I am forced to conclude that the trip will take 6 hours, regardless of whether I go today or tomorrow.

Second Student: Wait a minute! That answer is ridiculous! The presence or absence of a 5-mile-per-hour wind *must* make a difference to a trip in a sailboat. I agree with the first student's analysis of the trip if made today because the velocity of the wind relative to the water is what is important.

But if the trip were to be made tomorrow, under the "no wind" condition, then from the viewpoint of a boat drifting with the current at 5 miles per hour, through air that is motionless relative to the shore, the air acts as a 5-mile-per-hour headwind, blowing south to north. This headwind will impede the progress of the boat to some degree, even if the sail is lowered, so the journey would take a little longer tomorrow than today.

Professor: Your reasoning is excellent. Unfortunately, your knowledge of sailing is miniscule.

Suppose that you wanted to sail from north to south across a lake 30 ross, with no current. The wind is blowing from south to north. What would happen?

Third Student: I know. You would tack back and forth, in a zigzag path, and eventually you would reach your destination.

Professor: OK. Now, how would you apply that information to this problem?

Third Student: If I make my trip tomorrow, when there is no wind, I take advantage of the 5-mile-per-hour headwind created by the drifting of the boat. I tack back and forth as I drift downriver, and this maneuver adds to my speed. Therefore I get to my destination in less time than 6 hours. So if I don't mind a lot of tacking, I should make the journey tomorrow, not today.

1.3 FLOATING ON THE POND

Professor: This one is so simple that you should not need any time to think about it. Who has the answer?

First Student: Clearly, the canoe displaces a certain amount of water, depending on the weight of the canoe and its contents. When the rock is no longer in the canoe, the canoe will then displace less water, so the level of the pond will accordingly decrease. On the other hand, when the rock becomes immersed in the water, it must displace some water and accordingly causes an increase in the level of the pond. I would guess that these two effects would cancel one another. You haven't given us any quantitative information, so an exact comparison of the two effects is not possible.

Professor: Does someone have a different answer?

Second Student: The level of the pond will definitely decrease. As Archimedes first realized, the floating canoe displaces a volume of water which weighs exactly the same as the canoe and its contents.

Assume that the rock has a weight W_r, a volume V_r, and a density D_r, where D_r is greater than D_w, the density of water (since the rock sank to the bottom).

Removal of the rock causes a decrease in the volume of water displaced by the canoe by an amount equal to W_r/D_w.

Meanwhile the immersion of the rock in the pond displaces a volume of water equal to $V_r (= W_r / D_r)$. Since D_w is less than D_r, the net effect is a *decrease* in the volume displaced, which means a decrease in the water level.

If A is the surface area of the pond, the level will decrease by a distance equal to $W_r\,[(1/D_w) - (1/D_r)]/A$.

Professor: Very good. Now note that the preceding formula may be tested by assuming extreme conditions. If the rock were 100 times as dense as the water, clearly the change of displacement of the canoe would be the dominant effect rather than direct displacement of the water by the rock. For this case the formula is seen to reduce to $W_r\,A/D_w$.

At the other extreme, if the rock had the same density as the water, we would expect no net effect, a result that is confirmed by the formula.

1.4 THE BUBBLE IN THE WATER TANK

Professor: Here we have a bubble of air rising from a region where the absolute pressure is 3 atmospheres to a region where it is 2 atmospheres. According to the perfect gas law, the volume of the bubble must be inversely proportional to its pressure. Hence, the bubble must expand to 3/2 times its initial volume, as its pressure drops from 3 to 2 atmospheres.

But if the heavy-walled, steel tank is rigid and the water is incompressible, the bubble cannot expand. So what happens?

First Student: The perfect gas law involves absolute temperature as well as pressure. Therefore, if the absolute temperature of the air drops to two-thirds of its original value, no expansion is necessary.

Professor: True, but I have two objections to that explanation. First, by what physical mechanism would a decrease of temperature occur in a rising bubble of constant volume? Clearly, cooling by expansion cannot occur. Second, the water in the tank may be considered to be a constant-temperature reservoir, so the air bubble, surrounded by water, cannot change its temperature appreciably.

Second Student: We know that the bubble, once detached from the bottom of the tank, must rise to the top because air is so much lighter than water. We also know that the perfect gas law must hold. If we accept that the tank is rigid and the water is incompressible, we are forced to this paradoxical situation, in which the bubble, containing air at 3 atmospheres, is in contact with water at 2 atmospheres after rising to the top of the tank. This simply cannot be. Therefore, I must conclude that the tank is not quite rigid or the water is not quite compressible, so the bubble can expand.

Professor: The steel tank was said to be heavy-walled. Suppose the wall were a meter thick. The tensile strength of steel is great enough so that any expansion of the tank would be completely negligible for the pressures involved in this problem.

As for the compressibility of the water, the fact is that it can be compressed by 1 part in 20,000 by a 1-atmosphere increase in pressure. However we were not told what the volume of the bubble was relative to the volume of the water. If the bubble were very tiny, say 1/100,000 of the volume of the water, then water compressibility could play a role in the problem. But if the bubble were larger, say 1/1,000 of the volume of the water, then water compressibility clearly cannot explain the paradox.

Third Student: In view of these arguments, I am forced to conclude that unless it is a very small bubble, the rising of the bubble somehow causes the pressure of the water to increase everywhere in the tank, by 1 atmosphere. Then the new absolute pressure at the bottom is 4 instead of 3 atmospheres, and the new pressure at the top is 3 instead of 2 atmospheres. Thus, the bubble is still at 3 atmospheres and remains at constant size during and after rising.

Of course, this increase of pressure causes the diaphragm at the bottom of the tank to rupture since the pressure of 3.8 atmospheres (absolute) is exceeded. That was the original question.

Professor: So far, so good, as long as we are talking about a bubble that is large enough for the water to be considered incompressible. But we still want to understand the mechanism by which the pressure in the tank can rise.

Fourth Student: I propose this explanation for the mechanism by which the pressure can rise. Consider the same tank with *no* air bubble. However, imagine that a cylinder-piston arrangement with the same diameter as the formerly postulated bubble is present in the top surface of the tank, with the piston initially in contact with the water, which is at 2 atmospheres absolute pressure. Now suppose that an external force is applied to the piston, which causes it to exert a pressure of 3 atmospheres on the water. To the degree that the incompressibility assumption is valid, only a negligible motion of the piston will occur, but each point in the water will increase in pressure by 1 atmosphere.

Now, let us forget about the piston and visualize the original conditions, with the bubble at the bottom, containing air at 3 atmospheres. Then let us magically remove the bubble, replacing it with an equal volume of water, while simultaneously removing the same volume of water from the upper part of the tank and replacing it with a bubble containing air at 3 atmospheres. Obviously this bubble behaves in the same way as the piston, exerting pressure on the surrounding water and causing all the water in the tank to increase in pressure by 1 atmosphere.

Professor: That is a good explanation for the case of a sufficiently large bubble. We conclude that such a bubble will rupture the diaphragm, whereas a small bubble will simply expand while compressing the water slightly, producing a pressure rise too small to rupture the diaphragm.

Before leaving this problem, let us investigate the critical size of a bubble that will just rupture the diaphragm.

READER: Do you want to solve this problem? Then turn to page 12.

Professor: Let us define some symbols:

v_0 = initial volume of bubble
v_f = final volume of bubble
p_f = final pressure in bubble (absolute, in atmospheres)
V = volume of water in tank

We also note that each atmosphere increase in pressure causes a reduction in the volume of the water by 1 part in 20,000.

The following equation describes the compressibility effect:

$$(v_f - v_0) / V = (p_f - 2) / 20{,}000$$

This equation has two unknowns, v_f and p_f, so we need an additional equation. This equation comes from the perfect gas law:

$$p_f v_f = 3 v_0$$

If we eliminate v_f between these two equations and rearrange terms, we obtain

$$v_0 / V = [p_f - 2] / \{20{,}000[(3/p_f) - 1]\}$$

Now we know that when the pressure at the bottom of the tank is 3.8 atmospheres, the pressure at the top must be 1 atmosphere less, or 2.8 atmospheres (all absolute pressures). Accordingly, we substitute 2.8 for p_f in the preceding equation. We then find that $v_0 / V = 1/1786$. This is the critical volume of bubble, relative to the volume of water present, that will just rupture the diaphragm.

1.5 THE BOTTOM OF THE LAKE

Professor: Since your answer must be based on science and logic, not guesswork, how are you going to tackle this problem?

First Student: It seems to me that at the very least, you must give us local weather information for perhaps a year preceding next June 1. For example, what were the average daily temperatures? Also, what fraction of days had bright sun versus an overcast sun or precipitation? How much snow fell into the lake during the winter? Without such information, I see no basis for solving the problem.

Second Student: I don't need that type of information. I already know that Canada is cold in the winter, with substantial periods of subfreezing temperatures. That's all I need to know. I say that the temperature at the bottom of the lake next June 1 will be +4 degrees Celsius.

Professor: That is the right answer. Now will you please explain to the class how you obtained it?

Second Student: I happen to know that the density of water increases as it is cooled, until 4 degrees Celsius is reached, when the density reaches a maximum. On further cooling from 4 degrees down to zero degrees, the density decreases slightly. (See Figure 1.5A.)

Now let's consider what happens to this lake water as winter approaches. The surface water becomes cooler than the water beneath, and more dense, so it sinks down and is replaced by warmer water from below. This process continues until all the deep water is at the temperature of maximum density, 4 degrees C. Then further cooling of the surface water does not cause any more sinking. Ice forms on the surface and remains all winter. The water just below the ice is at zero degrees,

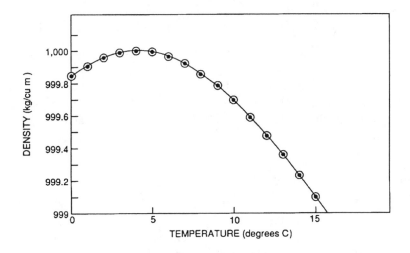

Figure 1.5A   Density of water versus temperature.

but the water 100 meters below the surface remains at 4 degrees. It would take many dozens of years for the water at the bottom to be cooled to zero degrees by thermal conduction through 100 meters of water.

When spring comes, the ice melts and the surface water warms up. However, the bottom water remains undisturbed, being at maximum density. On June 1, its temperature is still 4 degrees.

1.6 WHEN WILL THE PUDDLE EVAPORATE?

Professor: Does anyone feel that additional information is required?

First Student: I feel that the relative humidity in your dining room is an important variable. If the relative humidity is very low, say 10 percent, the evaporation will be much faster than if it is very high, say 80 percent. As a matter of fact, the ratio of the evaporation rates for these two conditions should be $(100 - 10)/(100 - 80)$, or 4.5. I say this because the relative humidity just above the surface must be 100 percent, and the difference between the vapor pressure at the surface and the vapor pressure well above the surface is the driving force for diffusion, which must control the rate of evaporation.

Furthermore, I obviously must know the temperature in the room since vapor pressure is highly dependent on temperature.

Professor: Good. Now does anyone see the need for more information?

Second Student: We do not know the size of the dining room or whether the doors and windows are open or shut. If the room is small and tightly sealed, the relative humidity in the room would increase as the water evaporates, and this condition would have to be taken into account. Indeed, it is conceivable that under some conditions 100 percent relative humidity would be reached before all the water had evaporated, so evaporation would cease at that time.

Professor: OK. Anything else?

Third Student: The rate of evaporation will depend on the motion of the air in the room. Perhaps there is a ventilation system or ceiling fan. Even if there is no mechanical ventilation, one or two of the walls of the room will probably be outside walls and will be at a slightly different temperature than the inside walls. This condition will cause convection currents in the room. I don't see how the problem can be solved until we know the velocity of the air just over the puddle.

Professor: I accept that requirement. If there are no more issues to be raised, I will supply values for the points you have mentioned, and then you will have 30 minutes to solve the problem, using engineering reference books as required.

The room temperature is 20 degrees C and the initial relative humidity is 40 percent. It is a closed room with a volume of 50 cubic meters. The air is flowing across the table at 1 meter per second.

READER: Try to solve the problem before turning to page 16.

Professor: Who has a solution?

Second Student: I calculated how much the relative humidity would increase in this closed room. My handbook tells me that saturated air at 20 degrees C contains 17.4 grams of water per cubic meter. So the air in the room, if saturated, would contain 17.4 x 50 = 870 grams of water. The air initially was at 40 percent relative humidity, so it contained 0.4 x 870 = 348 grams of water. I can now easily calculate that the addition of 100 grams of water would raise the relative humidity from 40 percent to 52 percent.

   However, on reflection, I realize that this is only an upper limit for the increase. The room probably contains furnishings, such as carpeting, drapes, upholstered chairs and almost certainly a ceiling and walls made of gypsum wallboard, and all these materials would have been initially equilibrated at 40 percent humidity. As the humidity rose during the evaporation, the exposed surfaces would absorb additional moisture from the air. Accordingly, the final value of humidity would be something less than 52 percent. Furthermore, I judge that at least several hours would be required before the water evaporated, and during this period there would probably be some leakage of air in and out of the room, even though it is nominally closed. You didn't say that it was perfectly sealed.

   In conclusion, I would say that a good estimate could be made of the *average* relative humidity during the evaporation period. It can be no more than 46 percent, which is halfway between 40 and 52 percent, and no less than 40 percent. If we assume that the average relative humidity during the evaporation period is 43 percent, this assumption could not possibly introduce an error larger than 7 percent.

Professor: That is a satisfactory estimate of the relative humidity of the air. But given that 43 percent relative humidity is a reasonable value to use, how do we proceed to estimate the time required for evaporation?

Third Student: I was able to find a formula for the thickness of a laminar boundary layer of air over a flat surface:

$$\delta = 1.3\sqrt{(L/V)} \qquad\qquad 1.1$$

where δ is the boundary layer thickness (cm), L is the distance from the leading edge of the flat surface (cm), and V is the velocity (cm/s). The water vapor must diffuse through this boundary layer to escape from the surface.

However, before I can calculate the boundary layer thickness from this formula, you have to tell me how close the puddle is to the edge of the table.

Professor: So you need more information after all. As you know, the boundary layer gets progressively thicker in the downstream direction. However, let's settle for the boundary layer thickness at the center of the pool. The center of the pool is 100 cm from the edge of the table.

Third Student: In that case, Equation 1.1 shows that the average boundary layer thickness is 1.3 centimeters.

Now, I can calculate the diffusion rate of water vapor across the boundary layer from the equation

$$r'' = D(c_{ws} - c_{w_\infty})/\delta \qquad\qquad 1.2$$

where r'' = mass flux of water vapor per unit area($g/cm^2 - s$)

D = diffusion coefficient (cm^2/s)

c_{ws} = concentration of water vapor at surface (g/cm^3)

c_{w_∞} = concentration of water vapor remote from surface

δ = average boundary layer thickness (cm)

We have already found that c_{ws} = 17.4 grams per cubic meter, which is 17.4 x 10^{-6} grams per cubic centimeter. Clearly, c_{w_∞} is 43 percent of this value, as was discussed. A handbook gives the diffusion coefficient of water vapor in air at 20 degrees C as 0.25 cm^2/s.

Upon substituting these values into Equation 1.2, we find that water is evaporating at a rate of 1.91 x 10^{-6} grams per square centimeter per second. Now, knowing the mass of water (100 grams) and the area of the puddle [π $(15)^2$ sq cm], we can easily calculate the time required, which comes out to be 74,100 seconds, or 20.6 hours.

Professor: So, we have an answer to the problem, which says that around 20 hours will be required. I now have to inform you that I carried out the experiment, making sure that the air velocity was 1 meter per second and the other quantities were as stated. I found that 32 hours were required, not 20 hours. This result suggests that we overlooked something significant. Who knows what it is?

READER: Do you know? Now turn to page 18.

Fourth Student: The difficulty is that we assumed the water to remain at the same temperature as the air, namely, 20 degrees C. Actually, as the water evaporates, the remaining water will become cooler. Accordingly, its vapor pressure will be reduced, and the rate of diffusion across the boundary layer will be correspondingly reduced.

A simple calculation will show the importance of this effect. My handbook says that the latent heat of vaporization of water is 2260 joules per gram, and the heat capacity of liquid water is 4.18 joules per gram per degree C. If the first 3 percent of the water evaporates, the heat of vaporization being supplied by the remaining 97 percent, then a simple heat balance shows that the temperature of the water must drop by

$$(3 \times 2260)/(9 \times 4.18) = 17 \text{ degrees C}$$

A temperature drop of 17 degrees, from 20 to 3 degrees C, would cause the vapor pressure to drop to 32 percent of its former value. Remember that the initial ambient relative humidity was 40 percent. Obviously, then, the temperature could not decrease to as low as 3 degrees C because evaporation would have ceased before then. However, we certainly cannot assume that the water remains at 20 degrees C.

Professor: Quite so. A handbook will show that at 5 degrees C, the vapor pressure of water will be the same as the partial pressure of water vapor in air at 20 degrees C and 40 percent relative humidity. Therefore, we know that the water temperature will decrease to some value between 5 and 20 degrees C, at which temperature a balance will exist between the rate of evaporative cooling and the rate of heat transfer from the surroundings to the puddle of water.

Before you can calculate the temperature at this balance point, you need to make some assumption in regard to the details of the heat transfer to the water. How do you think the heat is transferred?

READER: What do you think? Make a decision; then turn to page 20.

READER: *Turn page for solution.*

First Student: We have 100 grams of water spread out on a table top. The part of the table top directly below the water probably weighs a good deal more than 100 grams. As the water cools, it will pick up heat by conduction from the table top. We could calculate this heat transfer if we knew the thickness and thermal properties of the table top.

Second Student: I don't agree that this is the most important mode of heat transfer. Of course, if the table top were made of copper 3 inches thick, the first student would be correct. But it is more likely that the table top is not so massive. Notice that we are studying an evaporation process requiring something more than 20 hours. I can visualize that the table top under the water is being cooled during the first hour or so, but subsequently it will already have gotten nearly as cold as the water and will make little further contribution to the heat transfer.

It seems to me that a more important mode of heat transfer is the convective heat transfer from the air over the water, which will continue throughout the evaporation period. We already have calculated that the boundary layer thickness δ is 1.3 cm. The convective heat transfer per unit surface area of the water, q_c, is given by

$$q_c = k\Delta T/\delta \ watts/cm^2$$

1.3

where k is the thermal conductivity of air in the boundary layer and T is the temperature difference between the air and the water, which I assume to be constant after the first hour or so.

Professor: Wait a minute. Before you proceed to calculate with Equation 1.3, let's ask if anyone can visualize any other heat transfer mechanisms relevant to this problem.

Third Student: I remember that conduction, convection, and radiation are the three modes of heat transfer. We have mentioned conduction from the table top and convection from the air. What about radiation from the ceiling to the liquid puddle below?

Second Student: That's ridiculous! Radiation is only important when high temperatures are involved, which is not the case here.

Professor: Let's approach this problem objectively, by comparing q_r, the radiative heat flux per unit area, to the convective heat flux, q_c, which is given by Equation 1.3. The radiative exchange between the ceiling and the cooler evaporating water is given by

$$q_r = 5.67 \times 10^{-12} (293^4 - T^4) \; watts/cm^2 \qquad 1.4$$

where 5.67×10^{-12} is the Stefan-Boltzmann constant, 293 is the room temperature in Kelvins (273 + 20), and T is the water temperature in Kelvins. (We assume unit emissivity and absorptivity for the ceiling and the water.)

Since we do not yet know T, we should try to express Equation 1.4 in terms of ΔT. Then, when we take a ratio of Equation 1.4 to Equation 1.3, ΔT may be canceled. We rearrange the terms as follows:

$$293^4 - T^4 = 293^4 \{ 1 - [1 - (\Delta T/293)]^4 \}$$

where $\Delta T = 293 - T$, and use the approximation (valid if $\Delta T/293 << 1$):

$$[1 - (\Delta T/293)]^4 \approx 1 - (4\Delta T/293)$$

Then Equation 1.4 becomes

$$q_r \approx 4 \times 5.67 \times 10^{-12} \times 293^3 \times \Delta T \qquad 1.5$$

Now when we calculate q_r/q_c, from Equations 1.3 and 1.5, we find that ΔT is canceled:

$$q_r/q_c \approx 5.70 \times 10^{-4} \, \delta/k \qquad 1.6$$

From a handbook, k is 2.57×10^{-4} watts/cm$-K$ for air at 20 degrees C. Inserting this value and $\delta = 1.3$ cm into Equation 1.6, we find $q_r/q_c \approx 2.89$.

This calculation shows us that the radiative heat transfer from the ceiling provides nearly three times as much heat as the convective heat transfer from the air, under the conditions of this problem.

We may now obtain an equation for the total heat transfer (watts per sq cm) to the water by radiation and convection by summing Equations 1.5 and 1.3. After introducing values for k and δ, and expressing the water temperature as t (degrees C) instead of in Kelvins, we obtain

$$q \approx (5.70 \times 10^{-4} + 1.98 \times 10^{-4})(20 - t) = 7.68 \times 10^{-4}(20 - t) \quad 1.7$$

Now that we have Equation 1.7, how do we calculate the evaporation rate? We still haven't pinned down the temperature, t, of the water.

First Student: We know the heat of vaporization of the water, which is 2,260 joules per gram. If we divide Equation 1.7 by 2,260, we obtain an equation for the mass rate of evaporation of the water per square centimeter:

$$r'' = 3.40 \times 10^{-7}(20 - t)\ g/cm^2 - s \qquad 1.8$$

This formula tells us that the lower the temperature of the water, the greater is the rate of evaporation. Now recall Equation 1.2, with numerical values substituted for D, $c_{w\infty}$, and δ:

$$r'' = 0.19(c_{ws} - 7.5)g/cm^2 - s \qquad 1.2a$$

Here, c_{ws} is the concentration of water vapor just above the surface of the water, which is at temperature t. The lower this temperature, the lower this concentration will be, and accordingly, the lower the rate of evaporation.

Clearly, then, if we were to express c_{ws} in terms of t and then plot a graph of r'' versus t, with one curve dictated by Equation 1.8 and another curve dictated by Equation 1.2a, the two curves would cross at some temperature and some value of the evaporation rate. This is the solution to the problem.

Figure 1.6A is such a plot. The two curves are seen to intersect at 16.4 degrees C, and the corresponding evaporation rate is seen to be 1.23×10^{-6} $g/cm^2 - s$. Multiplying this value by the area of the puddle, 225π sq cm, we obtain $869 \times 10^{-6} g/s$ for the evaporation rate.

Therefore, the 100 grams of water will evaporate in $100/(869 \times 10^{-6}) =$ 115,000 seconds, or 32 hours.

Professor: Very good. Notice that the first (incorrect) calculation, which neglected the cooling of the water, gave 20.6 hours. We could do another (also incorrect) calculation, which neglects the radiative heat transfer but allows for the cooling of the water and its convective heating by the air, which would give us an answer of 60.6 hours. (The dashed line curve in the figure corresponds to this case.)

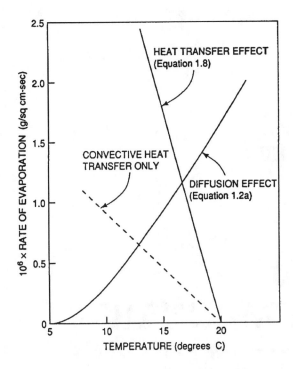

Figure 1.6A   Effects of heat transfer rate and diffusion rate on evaporation rate.

The lesson to be learned is that two people could each make an apparently logical analysis of this problem and get answers varying by a factor of three!

CHAPTER 2

SOME "ELEMENTARY" PROBLEMS

2.1 WHEN DO I ADD THE CREAM TO MY COFFEE?

I am served a cup of very hot, black coffee and a small container of cream, at room temperature. I want to drink the coffee (with cream) as soon as possible, once it has cooled to what I consider the proper temperature.

If I pour the cream into the coffee at once, I find that I then have to wait 5 minutes before it has cooled to the "proper temperature." I ask, "Suppose I had waited an interval, say 3 minutes, before pouring in the cream. Would the coffee have cooled to the proper temperature any sooner? Or does pouring in the cream at once result in the fastest cooling to the proper temperature?"

READER: Before turning to page 27, spend a little time thinking about this problem.

2.2 THE HOT-AIR BALLOON

I have a large, inflated balloon containing very hot air. It is completely sealed, except for a connection through which I can inject water. I inject some water at 25 degrees C, as a spray, and the water evaporates at once. See Figure 2.2A.

Does the balloon expand or contract?

READER: Provide your solution before turning to page 29.

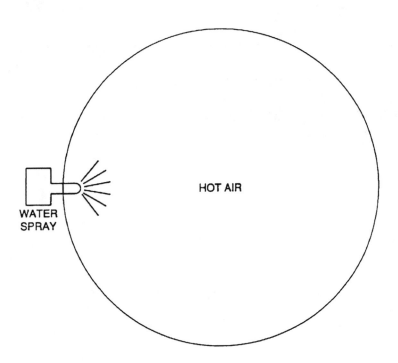

Figure 2.2A Hot-air balloon.

2.3 HOW FAR CAN AN AIRPLANE FLY?

Assume that I double the fuel load in a jet airplane by replacing part of the payload by fuel. The takeoff weight is unchanged. By what percentage will the range of the airplane increase?

Make simplifying assumptions as follows:

1. The airplane always flies at its design velocity, V.

2. Its lift coefficient, C_L and its drag coefficient, C_D, are independent of altitude. (These coefficients are proportionality constants between the lift or drag force per unit area and the dynamic pressure, $\rho V^2/2$.)

3. The fuel consumption rate is proportional to the thrust developed by the engines.

4. The total weight at takeoff is 100,000 kg, of which either 30,000 kg or 60,000 kg is fuel.

5. The flight altitude is controlled by the pilot to maximize the range.

6. The fuel consumed during takeoff, climb to altitude, descent, and landing is negligible compared with the fuel consumed when at altitude because it is a very long flight.

READER: Try to solve the problem, using all the simplifying assumptions, before turning to page 33.

SOLUTIONS TO CHAPTER 2
SOME "ELEMENTARY" PROBLEMS

2.1 WHEN DO I ADD THE CREAM TO MY COFFEE?

Professor: This problem is rather trivial, but the engineering principles involved in its analysis are instructive. What student has an idea of the solution?

First Student: I know that if I mix given quantities of a hot liquid and a cold liquid, the immediate reduction of temperature of the hot liquid will be greater the greater the temperature difference between the liquids. Accordingly, I say that you should pour in the cream at once to cool the coffee most rapidly to the so-called proper temperature. On the other hand, if you want the coffee to stay hot as long as possible, you should delay adding the cream.

Professor: Does anyone disagree with this analysis?

Second Student: The argument *sounds* plausible, but I think the problem is more complicated. The rate of cooling is not a constant but depends on the temperature of the coffee at each instant. I think the proper approach is to formulate the problem mathematically, invoking Newton's law of cooling, at least. We would need to know the initial temperatures, masses, and heat capacities of the coffee and cream. We also must know the ambient temperature, the "proper" temperature, and the heat transfer coefficient for heat loss from the coffee to the surroundings. Obviously, we must know the surface area of the coffee cup. If we want to solve the problem accurately, we should consider radiative and conductive heat loss, and possibly evaporative cooling, as well as convective heat loss. I doubt that we can be sure of the answer to the problem until we have calculated the thermal history of the coffee for both initial and delayed addition of the cream.

Professor: What you say makes sense, but such a calculation would be a lot of work for a "trivial" problem like this. Furthermore, it would only yield a specific rather than a general solution. Can anyone else suggest a different approach?

Third Student: I disagree with both the previous students. The first student reached the wrong conclusion. As for the second student's proposal, calculations are not necessary for this problem; only logic is needed.

The coffee, with the cream added, at its final, proper temperature clearly has a certain thermal energy, relative to room temperature. This is true regardless of *how* the coffee arrives at its final state. The thermal energy of the original very hot, black coffee is obviously larger than that of the final mixture, since we were told that a 5-minute cooling period was required after the cream was added. The quantity of heat that must be lost from the coffee or coffee-cream combination is *precisely* the difference between these two thermal energies, regardless of when the mixing occurs.

The problem comes down to this. How can we most rapidly lose the necessary quantity of thermal energy to the surroundings? Since the rate of heat loss by *any* mechanism (convection, conduction, radiation, evaporative cooling) is greater at any instant, the hotter the coffee is at that instant, it follows that I should delay as long as possible before adding the cream for most effective cooling. This must be true regardless of the ratio of the masses of coffee and cream or the initial temperature of the coffee or heat transfer parameters.

Professor: Congratulations. That sounds like excellent reasoning. However, your prescription, to delay as long as possible before adding the cream, would run into difficulty if you waited too long because then the result would be that the coffee with cream added would be below the desired proper temperature.

We know that the coffee will cool to the proper temperature in 5 minutes if we add the cream at once, so if we add the cream after waiting 5 minutes, your analysis shows that the coffee will then be below the desired temperature. Accordingly, the best time to add the cream is after a delay somewhat less than 5 minutes. This time can be determined either by calculation, as the previous student suggested, or by trial and error. If we try trial and error, a delay of 3 minutes would appear to be a good first trial. If the coffee is still too warm after the 3-minute delay, we need to wait only an additional short period that must be less than 2 minutes. If the coffee is too cold after the 3-minute delay, followed by the cream addition, we need to order a fresh cup of coffee and try a shorter delay.

2.2 THE HOT-AIR BALLOON

Professor: Who wishes to tackle this one?

First Student: It is immediately obvious that there are two opposing effects. The evaporation of the water will cool the hot air and tend to cause a contraction. On the other hand, we have increased the number of gas molecules in the balloon, which would tend to cause an expansion. Clearly a quantitative calculation is needed to see which effect dominates.

But we do not know the initial conditions (temperature, pressure, volume) of the hot air, nor do we know the quantity of water injected, so how can we make a calculation?

Professor: A *parametric* calculation can easily be performed once we identify the key parameters. One such parameter is the initial temperature of the hot air, which we will call T_1 and express in units of absolute temperature since we know we will use the perfect gas law. Another parameter is α, the molar ratio of water to air. (A molar ratio is more convenient than a weight ratio because the perfect gas law is easier to use with moles.) The absolute initial volume of the balloon is probably not relevant, as long as it is large enough so that no significant heat exchange occurs between the balloon material and the contents. It is obvious that the pressure is slightly above atmospheric. Accordingly, it is reasonable to try to calculate what happens when water is injected, in terms of the two parameters, T_1 and α.

First Student: I have another question. It is clear that the heat capacities of air and water vapor must be used in the calculation, and these are obtainable from engineering handbooks; but there are two kinds of heat capacities, those at constant pressure and those at constant volume. In this problem, if we assume the balloon expands, expansion must result from an increase of pressure, which increases the tension in the balloon membrane. Hence, both the internal pressure and the volume would have changed, and neither type of heat capacity seems to be correct.

Professor: A good question. However, we can bypass that difficulty as follows. Suppose we calculate whether the water injection would cause an increase or decrease of pressure in the balloon, at constant volume. Then we would know whether expansion or contraction would result.

Now I'll give you 5 minutes to develop your analysis, while I have a cup of coffee. You may want to know that the heat of vaporization of water at 25 degrees C is 44 kilojoules per mole, and the heat capacities at constant volume of steam and air at 100 degrees C are 25.9 and 21.0 joules/mole - K (Kelvins), respectively. These heat capacities vary somewhat with temperature, but you may wish to ignore this fact.

READER: You have a chance to improve your own analysis before turning to page 31.

Professor (after 5 minutes): Who has a solution?

Second Student: This is my analysis. I know that the heat of vaporization of the water is 44,000 joules per mole, at 298 K. If I assume that the final temperature, denoted T_2, is 500 K, then the sensible heat absorbed by the water vapor, at constant volume, is 25.9 x (500 - 298), or about 5,200 joules per mole. Since 5,200 is fairly small compared with 44,000, I can simplify the analysis by ignoring the sensible heat absorbed by the water and considering only the heat absorbed by vaporization. Then, my heat balance, for a constant-volume process, is

$$C_{air}(T_1 - T_2) = \alpha L \qquad 2.1$$

That is, the sensible heat removed by cooling a mole of air from T_1 to T_2 must equal the latent heat of vaporization of α moles of water.

We now use the perfect gas law, $PV = nRT$, where P is pressure, V is volume, n is number of moles, and R is the universal gas constant. It follows at once from this law that

$$P_2/P_1 = n_2\,T_2/(n_1\,T_1) = (1 + \alpha)T_2/T_1 \qquad 2.2$$

We know that T_2/T_1, the ratio of final to initial temperature of the balloon contents, is less than 1 since cooling is occurring. Also, we know that $1 + \alpha$ is greater than 1.

Thus, Equation 2.2 tells us that the pressure ratio can be either larger or smaller than 1, depending on the relative magnitudes of α and the temperature ratio.

We can rearrange Equation 2.1, solving it for T_2/T_1, and substitute the solution into Equation 2.2, obtaining

$$P_2/P_1 = 1 + \alpha - [\alpha(1 + \alpha)L]/[C_{air}T_1] \qquad 2.3$$

Equation 2.3 tells us what we want to know. The second term of the right-hand side is positive, whereas the third term is negative. Depending on the relative magnitudes of these two terms, we can see if the pressure rises or falls. More specifically, we see that if

$$T_1 < (1 + \alpha)L/C_{air}$$

the pressure must drop and the balloon must contract. We know that $L/C_{air} = 44,000/21 = 2,095$, so as long as the initial temperature is below $2,095 (1 + \alpha)$ K, the balloon contracts.

It is interesting to note that in the hypothetical case in which the initial temperature is well above 2,100 K and α is small, the analysis says that the balloon would expand.

Professor: An elegant analysis—except for your final statement. If the temperature were above 2,100 K, the sensible heat absorbed by the water vapor would be at least 25.9 (2,100 – 298) = 46,700 joules per mole of water, which is larger than the 44,000 value for the latent heat of vaporization. This result contradicts your assumption that the sensible heat was negligible compared with the latent heat.

Second Student: I agree. But even though I am now dubious that the crossover point between contraction and expansion is around 2,100 K, I feel that a more accurate analysis would reveal a crossover point at some temperature for any small value of α.

Professor: Has anyone else done an analysis that sheds light on this possible crossover point?

Third Student: I have. First, I retained both the C_{air} and the C_{vap} terms in the energy balance, but the equations became more complicated than I liked. Then, noting that the heat capacity values for air and water vapor, 21.0 and 25.9, were not very far apart, I decided to pretend that they were both the same. I assumed that both air and water vapor had heat capacities at constant volume of 23 joules per mole per K, and I denoted this heat capacity as C. Then the equations became simple.

The energy balance becomes

$$C(T_1 - T_2) = \alpha[L + C (T_2 - 298)] \qquad\qquad 2.4$$

and the pressure ratio equation is the same as that of the previous student (Equation 2.1). Combining these two equations yields

$$P_2/P_1 = 1 - [(L/C) - 298]\alpha/T_1 \qquad\qquad 2.5$$

Equation 2.5 says that the pressure ratio is always less than 1 as long as L/C is larger than 298, regardless of the values of T_1 and α. I calculate that

L/C = 44,000/23 = 1,913, which is far larger than 298. Therefore I conclude that the balloon always contracts.

Professor: I like your analysis better than the previous one because your assumption of equal heat capacities for air and water vapor is closer to reality than the assumption of neglecting the heat capacity of water vapor entirely.

However, both treatments fail to consider the variation of heat capacity with temperature. This would not matter much as long as $T_1 - 298$ was no more than a few hundred degrees. But the heat capacity at constant volume of water vapor is 1.72 times as great at 2,100 K as at 298 K, and we could not be sure what would happen at such extreme temperatures without doing detailed calculations not involving approximations.

I have performed such a calculation, selecting $T_1 = 2,100\ K$ and $\alpha = 0.01$. T_2 came out to be 2,062 K, and the pressure decreased 0.8 percent. This calculation took into account the variations of heat capacities with temperature (but ignored dissociation of water vapor). Surprisingly, the result agreed very well with the prediction of Equation 2.5. At even higher temperatures, dissociation of water vapor would become important, and more elaborate calculation would be needed.

Of course, we are ignoring the practical fact that no balloon could survive at such high temperatures.

2.3 HOW FAR CAN AN AIRPLANE FLY?

Professor: Who knows how to solve this problem?

First Student: The problem seems to be very easy, in view of all the simplifying assumptions. Obviously, if the fuel load is doubled, everything else being the same, the range is doubled.

Professor: But the problem never stated that everything else is the same. If the plane always flew at the same altitude (selected by some optimization procedure), your solution would be correct. However, according to the assumptions, the pilot is free to adjust the altitude continuously to maximize the range.

Let's look into this more closely. We know that the density, ρ, of the atmosphere decreases nearly exponentially from the sea-level density, ρ_s as the altitude, h (kilometers), increases, according to the relationship

$$ln\,(\rho_s/\rho) = 0.104h \qquad\qquad 2.6$$

Since the drag is proportional to the density, ρ, the pilot wants to fly as high as possible to minimize drag.

However, the higher the altitude, the less the lift that the wings can produce at the given flight velocity, V. The lift cannot be less than the weight of the airplane, or level flight cannot be sustained. The weight of the airplane decreases as fuel is consumed. Accordingly, to maximize the range, the pilot should gradually increase the altitude as fuel is consumed and the airplane becomes lighter.

Consider now the flight profiles for the two cases (30,000 and 60,000 kg of fuel). The initial weight of the airplane is the same in both cases, so the initial optimum altitude is the same. If we were to plot the optimum altitude versus the instantaneous weight of the airplane, the plots would be exactly the same for the two cases until the first 30,000 kg of fuel were consumed in each case. In the first case, the plane would have to land at this time, whereas in the second case a relatively light plane would start the second segment of its flight, which would be accomplished at higher altitude. Accordingly, the air resistance and the necessary rate of fuel consumption to maintain the speed V would be less, and the airplane would be able to fly longer and go significantly further during this second segment of the flight than in the first segment.

Clearly, then, doubling the fuel load much more than doubles the range.

First Student: I follow your argument. But we still need to calculate *how much* extra range we get in case 2. How do we do that?

Professor: We must develop some equations. We know that in steady flight, the aerodynamic drag on the airplane must equal the thrust of the engines, which is proportional to the fuel consumption rate. We also know that the aerodynamic lift must balance the weight of the airplane at each instant. Drag and lift are described by equations involving drag coefficient, C_D, and lift coefficient, C_L:

$$\text{Drag force} = C_D\,A_D\,\rho\,V^2/2 = \text{thrust force} = k(-dM_F/dt) \qquad 2.7$$

$$\text{Lift force} = C_L\,A_L\,\rho V^2/2 = (M_D + M_F)g \qquad\qquad 2.8$$

In these equations, A_D is the area of the airplane responsible for drag (see note on p. 38), A_L is the area producing lift, M_D is the "dry" mass of the airplane, M_F is the mass of fuel aboard at any time, k is a measure of the engine efficiency, t is time, and g is the local acceleration due to gravity. (It is assumed that the SI system of units is being used.)

Now, if we divide Equation 2.7 by Equation 2.8 and rearrange, we obtain

$$dM_F/(M_D + M_F) = -[(C_D A_D g)/(C_L A_L k)]dt = -Kdt \qquad 2.9$$

The quantities in the square brackets are all constants in this problem and may be replaced with a single constant, K. Equation 2.9 is seen to be a simple differential equation, which may be integrated by using limits as follows. The initial condition is $M_F = M_{F0}$ and $t = 0$; while the final condition is $M_F = 0$ and $t = t_f$.

The result of the integration, after the limits are applied, is

$$ln[(M_D + M_{F0})/M_D] = Kt_f \qquad 2.10$$

but the range of the airplane is given by

$$\text{Range} = Vt_f \qquad 2.11$$

Hence, from Equations 2.10 and 2.11, the range is

$$\text{Range} = (V/K)ln(\text{initial mass/final mass}) \qquad 2.12$$

It is now easy to calculate that the ratio of the ranges for the two cases (60 percent and 30 percent initial fuel) is

$$ln(100/40)/ln(100/70) = 2.57$$

Your original assumption that the range doubles when the fuel load doubles is seen to be a substantial underestimate! In fact, a 100 percent increase in fuel gives a 157 percent increase in range.

Second Student: I follow your analysis, but I still have a question. Is it possible to calculate the optimum flight profile from the preceding information?

Professor: Yes, of course. It is actually very simple. All I have to do is to combine Equations 2.6, 2.8, and 2.10, remembering that $lnA + lnB = ln(AB)$. For the optimum height during the initial part of the flight, which I call h_0, I obtain the relation

$$h_0 = (1/0.104)ln \left\{ (\rho_s C_L A_L V^2)/[2g(M_{F0} + M_D)] \right\} \qquad 2.13$$

For time greater than zero, I obtain

$$h = h_0 + Kt \qquad 2.14$$

where K was defined in Equation 2.9 as a group of constants. Equation 2.14 shows a simple linear relation between flight altitude and time.

Finally, it is interesting to note that the dependence of the range on the logarithm of the ratio of initial to final mass, which we found, has absolutely nothing to do with the logarithmic dependence of air density on altitude. If you look back over the equations, this will be clear.

Footnote: In the foregoing analysis, the drag was expressed as $C_D A_D \rho V^2/2$. Actually, there are three types of drag affecting airplanes: form drag, skin drag, and drag associated with shock waves. In this case we assume that the flight velocity is well below the speed of sound, so we ignore shock waves. The form drag involves the cross-sectional area of the airplane normal to the flight direction, whereas the skin drag involves the surfaces of the airplane past which the air flows, such as the wings. A separate drag equation could be used for each of these forms of drag and summed. In this problem, we ignore these complications. Also we ignore the small variations of lift coefficient, C_L, and engine efficiency factor, k, with altitude, so as not to obscure the simple (first approximation) solution, as shown by Equation 2.12, which is known as the Breguet range equation.

CHAPTER 3

WHAT IS THE CHANCE THAT . . . ?

3.1 INTRODUCTION

We are all aware that precise probability calculations can be made in many cases. For example, the probability that a spinning coin will stop in the heads rather than the tails position is clearly 0.5. As a slightly more complicated case, if I am playing bridge, and if my side has 8 of the 13 hearts, the two opponents having the remaining 5, I can calculate the probability that those 5 cards are split 5 to zero, 4 to 1, or 3 to 2. (These probabilities are 0.0625, 0.3125, and 0.625, respectively, assuming random dealing.)

However, when we attempt to apply probability to real-life situations, which are often more complex than coins, dice, or playing cards, we find that because of subtle effects, the calculations are not so simple.

Let's imagine that I live in a city with 100,000 houses. Suppose that I read in the newspaper that a resident of my city, who owned a small private airplane, crashed into the roof of his own house while flying on Sunday afternoon.

Among the 100,000 houses, what is the probability that he would have crashed into his own house?

At this point, you are surely thinking that the pilot must have deliberately flown over his own house at a low altitude. Possibly his motive was to make an impression on his own family. Possibly he wanted to make a visual inspection of his roof because of leakage problems. Conceivably, he had an urge to commit suicide in this bizarre fashion. Whatever his reason, however, you will agree that the presence of his airplane in close proximity to his house was almost certainly the result of deliberate action on his part and should not be treated as a random event.

Accordingly, if I were to assert that the probability of his crashing into his own house instead of some other house was 1 in 100,000, I would expect you to reject this assertion as meaningless.

Let us now turn from this little example to the general question of assessing probabilities. To perform a risk analysis for a practical problem, such as the siting of a nuclear reactor or a liquefied natural gas storage tank or a choice between alternate methods of travel to a given spot, one must assign numerical values to probabilities.

In general, this task is done when possible by studying statistics relating to failure modes for similar situations from the past. The approach is to postulate all conceivable failure modes and try to associate probabilities with each. If two independent events both must occur to produce the failure, one uses the principle that the probability of the failure is equal to the product of the probabilities of the two events.

In making such evaluations of probabilities, it is necessary to proceed with extreme caution since, otherwise, gross errors can result. Are the two "independent" events really independent? Is the 1 house out of 100,000 truly selected "at random," or are there hidden factors at work?

In the following sections are some examples of ways in which one could easily go wrong in calculating probabilities.

3.2 DIVISION BY 137

What is the probability that the prime number 137 is a factor of an 8-digit number selected "at random"?

Professor: Who wants to answer this question?

First Student: It's embarrassingly simple. If we write all the digits from 1 to infinity in sequence, we will find that 137 will divide exactly into 137, into 2×137, into 3×137, and so forth. In short, it will divide exactly into every 137th number and into no intervening numbers. Obviously, the answer to the question is 1/137.

Professor: Any other comments?

Second Student: I am intrigued by the quotation marks around the phrase "at random" in the question. Exactly how is this 8-digit number to be selected?

Professor: That is an interesting question. It happens that I asked a computer programmer how to obtain a group of 8-digit numbers so that I could test them, dividing each by 137 to solve the problem experimentally.

He told me that he will write a computer program that will produce the random numbers and solve the problem for me. Unfortunately, however, his computer can only produce 4-digit random numbers. Therefore, he decided to generate 8-digit random numbers by using each 4-digit random number twice. That is, the 4-digit number, *abcd*, is converted into an 8-digit number, *abcdabcd*. Having generated 10,000 such 8-digit numbers, his computer program divided each of them by 137 and noted if there was a remainder.

READER: What do you think was the result? And why? Then turn to page 41 for the solution.

3.3 THE TURBINE SEALS

Two steam turbines are located in a paper mill and generate electricity to operate the plant. The turbines are designated number 1 and number 2. Once a year, the mill is shut down for a week while repair and maintenance chores are performed. As part of this procedure, the two turbines are partially disassembled, carefully inspected, and then reassembled.

Four days after this maintenance period the plant is back in operation. The supervisor of the turbine room says to his assistant, "Joe, when you replaced the seal disc on each of the turbines last week, you did make absolutely sure that the notched side of the disc was facing outward, didn't you?"

Joe responds, "No one told me anything about that. I replaced the seal discs on turbine number 1 and turbine number 2, but I have no idea which way either one is facing."

The supervisor is alarmed because incorrect orientation of the seal may ultimately cause turbine damage. However, inspection of the seals is not now possible without interfering with plant production. The turbine supervisor realizes that he must discuss the situation with the plant manager.

To prepare for this discussion, the supervisor feels that he should first evaluate the probabilities that one or both seals are incorrectly oriented.

At this point, the assistant says, "Wait a minute. We may have some additional helpful information. Just before reassembly of the turbines last week, we took some photographs. Possibly, they may show how the seals are oriented."

The photographs are then studied, and, sure enough, a seal is clearly visible in one of the shots and it is correctly oriented. Unfortunately, it is not possible to tell if this is a picture of turbine number 1 or turbine number 2.

From these facts, the supervisor now calculates the probability that a problem exists and then goes to see the plant manager.

READER: How would you evaluate the probability? Then turn the page 42 for the solution.

3.4 THE KEYS TO THE TAXICABS

I am starting a company to provide taxicab service. I place an order for 20 new taxicabs. I tell the salesperson that I want all 20 ignition keys to be different from one another.

The salesperson replies, "We have 200 different locks we can use, with corresponding keys. As each taxicab goes through our computer-controlled assembly line, one of these 200 types of locks is selected at random for installation. Accordingly, the probability that the key for any taxicab will fit any other taxicab is only 1/200.

"If you specify that all 20 keys *must* be different for your fleet, there would be a surcharge of $300. On the other hand, if you take the 20 cabs as they come, you could then, if necessary, have one of those locks changed for a charge of $500. But, most likely, this would not be necessary.

READER: What is the optimum course of action? When you are ready, turn to page 42 for the solution.

SOLUTIONS TO CHAPTER 3
WHAT IS THE CHANCE THAT...?

3.2 DIVISION BY 137

Professor: To continue the story, he found that there was no remainder in any one of the 10,000 cases he tried. The apparent conclusion is that the probability in question is unity instead of 1/137. What do you think about that?

First Student: I am satisfied that the correct probability is 1/137, and it is clear that his method of generating 8-digit random numbers is flawed. However, I am at a loss to explain how every number he tested was divisible by 137, unless there was some sort of error in his computer program.

Professor: What are the last four digits in your home telephone number?

First Student: 3290.

Professor: When you divide 32903290 by 137, is there a remainder?

First Student (after one minute): It divides exactly. No remainder.

READER: Try it with *your* phone number.

Professor: Will you all take 10 minutes to think about this situation, and then provide the explanation?

Second Student (ten minutes later): I have an explanation. If I multiply any 4-digit number, *abcd*, by the number 10,001, the product is *abcdabcd*. Hence 10,001 is a factor of any number of the form *abcdabcd*. But 10,001 is not a prime number; instead, it has two prime factors, 137 and 73. Accordingly, if I divide 137 into *abcdabcd*, the quotient is 73 x *abcd*, and there is no remainder.

Professor: Very good. However, you students may feel that this was a trick problem and not representative of what one might encounter in the real world. The next two problems are somewhat more realistic.

3.3 THE TURBINE SEALS

Professor: You now have all the facts that the supervisor has. How do you evaluate the probability that one of the seals is incorrectly installed?

First Student: Before we heard about the photographs, we could visualize four equally likely possibilities:

- Seal number 1 OK; seal number 2 OK
- Seal number 1 OK; seal number 2 wrong
- Seal number 1 wrong; seal number 2 OK
- Seal number 1 wrong; seal number 2 wrong.

Therefore, there is a 25 percent chance of each of these possibilities, and hence, a 75 percent chance that something is wrong.

However, once we know about the photograph, the fourth possibility in the list (both seals wrong) can be eliminated. Looking at the remaining three possibilities, I see that there is a 66.6 percent chance that something is wrong.

Professor: Does anyone disagree with that analysis?

Second Student: I do. If the photograph showing the correctly placed seal is a photograph of turbine number 1, there is a 50 percent chance that the seal in turbine 2 is wrong. On the other hand, if it is a photograph of turbine number 2, there is a 50 percent chance that the seal in turbine number 1 is wrong. Either way, the probability of an incorrectly oriented seal is 50 percent.

Professor: You are right. But what was wrong with the first student's analysis?

Third Student: Before the examination of the photograph, the four enumerated possibilities were all equally probable. After the photograph was examined, it is true that the fourth possibility is eliminated, but it is not true that the remaining three possibilities remain equally probable. Once we know that at least one seal is correct, the first possibility (both seals correct) is clearly as probable as the sum of the probabilities of the second and third possibilities (one seal right and one seal wrong).

3.4 THE KEYS TO THE TAXICABS

Professor: You have had an hour to study this problem. Now, what is your recommended course of action?

First Student: In view of my past record of being consistently wrong, I don't have a lot of faith in my answer. But here it is.

I assume that the 20 cabs are sitting on the parking lot. I select any one at random, and try to fit its key into each of the other 19 cabs. In each case I have a 1/200 probability of a fit. Therefore I have a 19/200, or 9.5 percent, probability that the randomly selected key will fit another cab.

A 9.5 percent probability of a $500 expenditure gives a probable cost of $47.50. This is obviously preferable to a 100 percent certainty of a $300 cost. So I would take the cars as produced on the assembly line.

Professor: I am sorry, but your solution is wrong. Even if the randomly selected cab key doesn't fit any of the other cabs, there is no assurance that some of the other cab keys may not fit one another's locks.

Who has another solution?

Second Student: Here is my analysis. The cars come off the assembly line one at a time. Let us number them 1 to 20. The probability that the key of cab number 2 will fit cab number 1 is 1/200. As for cab number 3, its key has a 1/200 chance of fitting cab number 1 and also a 1/200 chance of fitting cab number 2. Hence, its key has a 2/200 chance of fitting a preceding cab.

By the same reasoning, the key of cab number 4 has a 3/200 chance of fitting a previous cab; generalizing, the nth cab's key has a $(n-1)/200$ chance of fitting a previous cab.

The overall probability that some key will fit a different cab is simply the sum of these probabilities:

$$1/200 + 2/200 + 3/200 + \ldots + 19/200 = 0.95.$$

Hence, it is better to spend the $300 to ensure no duplication rather than having a 95 percent probability of spending $500.

Professor: Sorry, but you missed the boat also. Suppose you had been purchasing 21 cars instead of 20. Then your logic would have been the same, but your series would have had an additional term, 20/200, and the sum would have been 1.05. But a probability greater than unity is clearly impossible. In fact, it is easy to see that with 200 different kinds of keys, you would have to purchase at least 201 taxicabs to have a certainty of a duplication. So your reasoning must contain a fallacy. Who sees what it is?

Third Student: Let's consider the first three cars off the assembly line. The key of car number 3 has a 1/200 chance of fitting car number 1. *If we* *knew* that car number 2 had a different lock from car number 1, we could say that the chance that key number 3 fits either car number 1 or car number 2 is exactly twice as great as 1/200. But we have no such assurance, so the analysis is not strictly correct. Of course, it would be very close to correct if we were dealing with only three cars, but it becomes increasingly wrong as we consider larger numbers of cars.

This is the way I analyzed the problem. Let's stay with the idea of the cars being delivered in sequence. The probability that the second cab key will *not* fit the first cab is 199/200. In other words, there is a 199/200 probability that the first two cab locks are different.

Whenever that condition exists, there is a 1/200 chance that the third cab key will fit the first cab and also a 1/200 chance that it will fit the second cab; hence there is a 198/200 chance that the third key will not fit either the first or second cab when the first two cab locks are different (which has a likelihood of 199/200). If we multiply these two probabilities together, we obtain the probability that the first three cab locks are all different from one another. This is

$$(199/200) \times (198/200)$$

By exactly the same reasoning, the probability that all keys are different for four cabs is

$$(199/200) \times (198/200) \times (197/200)$$

And finally, for 20 cabs, the probability of no interchangeable key is

$$(199/200) \times (198/200) \times (197/200) \times \ldots \times (181/200) = 0.374$$

Accordingly, the probability of at least one case of interchangeability is 1 minus 0.374, or 0.626. If there is no more than one case of interchangeability, the expectation of expenditure is 0.626 x $500, or $313. However we might have to change not one but two or three locks, at a greater cost. There is no need to calculate this figure, however, since the best strategy is to pay the $300 surcharge.

Professor: You are quite correct. However, notice that the previous two analyses, although sounding plausible, gave grossly different results. Clearly, great caution is needed in making such calculations.

CHAPTER 4

UNSTABLE BEHAVIOR

4.1 INTRODUCTION

Unstable behavior can occur not only in mechanical arrangements but also in a wide range of physical and chemical processes involving unstable temperature or pressure. Even in pure mathematics, it is found that certain simple equations have unstable solutions. Whenever instability occurs, unexpected events may result.

A very simple example of mathematical instability may be illustrated with the iterative equation

$$X_{n+1} = CX_n(1 - X_n)$$

If we assign a value to the constant C, say $C = 2.8$, and then select an initial value for X, say $X = 0.1$, and then proceed to calculate successive iterations of the equation, we would obtain the result shown in Figure 4.1A. The values of X_{n+1} oscillate but converge fairly rapidly to a value of 0.643.

If we were to repeat this procedure but start with a different initial value, say $X = 0.3$ instead of 0.1, the X values would again converge to the same limit, 0.643.

There is nothing so remarkable about any of this discussion. But if we select a different value for the constant C, say 3.6 instead of 2.8, and again calculate successive iterations, starting with $X = 0.1$, the results would be

45

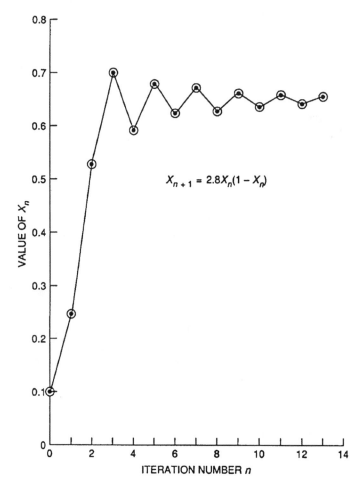

Figure 4.1A   Converging iteration.

as shown in Figure 4.1B. It is seen that the X values wander up and down in a chaotic way with no tendency to converge to any final value.

Obviously, when the C value in the equation is changed from 2.8 to 3.6, some sort of mathematical instability is introduced. This phenomenom is very interesting and is an important subject of mathematical research. But our concern is with unstable phenomena in the real world.

A mechanical system is stable when it responds to a small perturba-

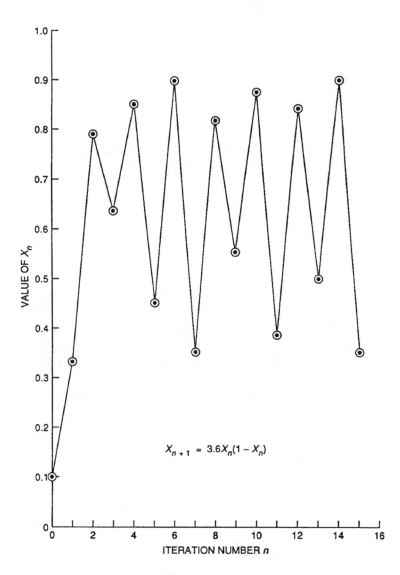

$$X_{n+1} = 3.6X_n(1 - X_n)$$

Figure 4.1B Non-converging iteration.

tion by returning fairly soon to its original position. Figure 4.1C shows three classic cases. In the first case (stable), the sphere, if disturbed, will return to its original position after some oscillations damped by friction. In the second case (unstable), the sphere, if disturbed by even a very small force, will never return to the initial condition. In the third case

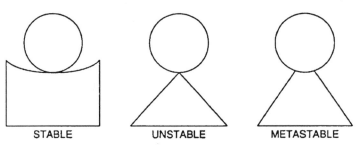

STABLE   UNSTABLE   METASTABLE

Figure 4.1C   Stability.

(metastable), the sphere is stable to a small perturbation but unstable to a larger perturbation.

A less obvious case of mechanical stability is that of a rolling disc. Imagine a dinner plate rolling on its rim on a flat floor. As long as its speed is greater than a certain minimum, it will be stable, but as it slows down its speed will eventually decrease below the necessary minimum and the plate will topple over. The critical speed is the square root of $gd/6$, where g is the acceleration of gravity and d is the diameter of the disc. For a disc 1 meter in diameter, the critical speed comes out to be 1.28 meters per second, or 2.86 miles per hour. (By this same principle it is possible to ride a bicycle, but not very slowly.)

After these introductory remarks on stability, let's look at the problems that follow.

4.2 THE SOLID-PROPELLANT ROCKET

Consider a steadily burning solid-propellant rocket. The higher the pressure in the combustion chamber, the faster the propellant burns. (The main reason for this relationship is that the higher the pressure, the closer the gaseous flame is to the surface and the more effectively it can transfer heat back to the surface, thereby gasifying the propellant more rapidly.) The rate of gas generation, r_g, may be described by the equation

$$r_g = C_1 P^n$$

where P is the pressure, and C_1 and n are constants for a given propellant with given exposed surface area.

The hot gas leaving the rocket must pass through the throat of the exhaust nozzle at sonic velocity. The higher the pressure, the greater is the density of this gas and the greater the rate of mass flow at sonic velocity. (The speed of sound in a gas is independent of pressure.) Accordingly, the rate of gas escape, r_e, may be described by the equation

$$r_e = C_2 P$$

C_2 is a constant, dependent on the cross-sectional area of the nozzle throat and on the velocity of sound in the combustion products.

What is the steady operating pressure of a solid-propellant rocket for given values of the constants C_1, C_2, and n?

READER: Can you answer this question, keeping in mind that this problem is in a section on unstable behavior? Then turn to page 51.

4.3 BOILING WATER WITH A HEATED WIRE

An aluminum wire, immersed in a pool of water at 100 degrees C, is electrically heated by passing a large current through it. Boiling occurs along the wire. The electric voltage applied to the wire is gradually increased. The wire gets progressively hotter and the boiling rate increases.

At a given time, the wire is found to be at 140 degrees C. (The melting temperature of aluminum is 660 degrees C.) The voltage is then increased another 2 percent, and the wire promptly melts.

What kind of instability is this?

READER: Do you understand what occurred? Now turn to page 54 for the solution.

4.4 A CATALYTIC PARTICLE

We have an exothermically reactive gas mixture at temperature T_g. However the mixture is so diluted with excess air that it is nonflammable (e.g., 1 percent hydrogen in air).

Suspended in the gas mixture is a small catalyst particle (e.g., platinum), which is initially at temperature T_{p0}. The gas reacts exothermically on the surface of the particle, so its temperature rises. Once a steady state is reached, the particle will be at a temperature T_p.

How would we go about calculating this temperature? What additional facts would we need?

READER: What is your approach to this problem? Then turn to page 56 for the solution.

SOLUTIONS TO CHAPTER 4
UNSTABLE BEHAVIOR

4.2 THE SOLID-PROPELLANT ROCKET

Professor: Doesn't this sound like an easy problem? Who has the solution?

First Student: At the steady state, which we are seeking, the pressure in the combustion chamber must adjust itself so that the rate of gas generation is equal to the rate of gas escape. Equating the two rates, we have

$$C_1 P^n = C_2 P$$

Solving for P, we obtain

$$P = (C_1/C_2)^{1/(1-n)}$$

This is the answer.

Professor: Your algebra is impeccable, but let's look at the physics involved. Figure 4.2A is a plot of the gas-generation curve and the gas-escape curve, with n taken to be 1.3. The intersection of the two curves, which you presume to be the steady-state solution, is right where it should be, according to your formula.

But what about stability? If you look at the region around the intersection of the two curves, you see that if the pressure should momentarily be a fraction higher than the intersection value, the rate of gas generation exceeds the rate of gas escape, and the pressure rises further. This occurrence causes an additional increase in the rate of generation, and in a very short time the pressure rises to the point where the rocket explodes.

On the other hand, if the pressure should momentarily be a fraction

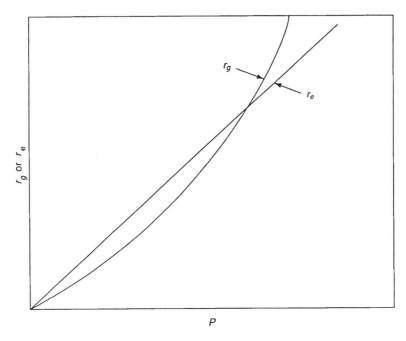

Figure 4.2A Rates of gas generation, r_g, and gas escape, r_e, versus pressure, P. The intersection is unstable.

lower than the intersection value, the rate of generation is less than the rate of escape, and the pressure drops further. This result causes flameout in a very short time.

Thus, your solution appears to be mathematically correct but physically incorrect. How, then, can a solid-propellant rocket work?

Second Student: Professor, your argument was based on a value of n greater than 1, namely 1.3. Suppose that n was less than 1, for example, 0.8. Then the situation would be entirely different (see Figure 4.2B). If the pressure were momentarily a little higher than at the intersection of the two curves, the rate of gas escape would be greater than the rate of gas generation, and the pressure would drop back to the equilibrium point. Similarly, if it dropped to a value a little lower than the equilibrium point, the rate of gas generation would then be higher than the rate

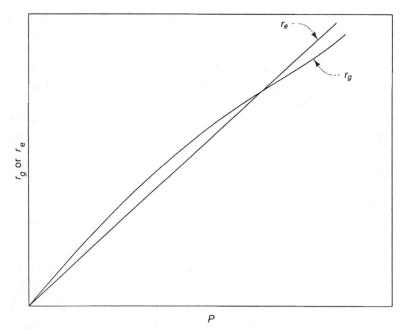

Figure 4.2B Rates of gas generation, r_g, and gas escape, r_e, versus pressure, P. The intersection is stable.

of escape, and the pressure would rise back to the intersection value. Thus, stability!

I conclude that the value of n for a solid propellant must be less than 1 or stable burning in the rocket cannot occur.

Finally, I note that the slope of the gas-generation curve at the intersection is clearly smaller than the slope of the gas-escape curve, as long as n is less than 1. This comparison of slopes gives us a criterion of stability.

Professor: Very good. But a question remains. Suppose that an analyst had derived the formula for the equilibrium pressure:

$$P = (C_1/C_2)^{1/(1-n)}$$

What clue would the analyst have from looking at this formula that there was a need to investigate stability?

Third Student: The most casual inspection of the formula shows that when $n = 1$, P must be infinite. This finding calls for a more careful look at the formula. We recall that the constant C_1 is proportional to the rate of gas generation. We see from the formula that the equilibrium pressure increases when C_1 increases, as long as n is less than 1, which is physically reasonable. Furthermore, we see that for values of n greater than 1, the formula says that the equilibrium pressure is an *inverse* function of C_1, which is not physically reasonable. The formula is trying to talk to us and provide clues that there is a problem when n equals or exceeds unity.

4.3 BOILING WATER WITH A HEATED WIRE

Professor: If you have studied heat transfer, you already know the answer to this question. Will someone provide it?

First Student: It turns out that I am going to get at least one answer correct. When the wire is only moderately hotter than the surrounding water, which is 100 degrees C, the mode of boiling, called nucleate boiling, consists of tiny steam bubbles forming at various points on the wire (nuclei), growing in size, and then detaching themselves. They then rise to the surface and are replaced by as yet unboiled water. This is a very efficient mode of heat transfer.

However, if the wire is quite hot relative to the water, a film of steam forms immediately around the wire, like a sleeve. Since the thermal conductivity of steam is much less than that of water, the steam acts as insulation, and further heat transfer is quite slow. If electric energy continues to be supplied to the wire, it will get hotter and hotter, and ultimately the temperature gradient across the steam film will get large enough to reach a steady state. In this steady state, steam will occasionally detach itself from the film and rise to the surface, but a new film will promptly form over the surface, so the wire is almost never in contact with the liquid water. This mode of boiling is called film boiling and is characterized by very high wire temperatures.

Of course, if the temperature corresponding to the steady-state film-boiling condition is higher than the melting point of the wire, as is the case in this problem, the wire melts when the transition from nucleate to film boiling takes place.

Professor: Excellent. I have prepared a graph (Figure 4.3A) showing the steady-state heat flux emerging from the surface of a heated wire versus the excess temperature of the wire relative to the surrounding water. As the electric power input is increased, the temperature of the wire rises, reaching a critical point, marked A on the curve, when the heat flux is about 300 watts per square centimeter and the excess temperature is about 40 degrees C (140 − 100). Any further increase in heat flux leads to an unstable condition, and the wire jumps to a temperature of over 1,000 degrees C, or in this case, it melts at 660 degrees C.

Now, I want you to tell me just why the region on the curve between A and C is unstable.

Second Student: First, let's talk about the region between A and B, which shows increasing temperature and decreasing heat flux. This behavior is physically reasonable because we are in a transitional region, changing from efficient nucleate-boiling heat transfer to inefficient

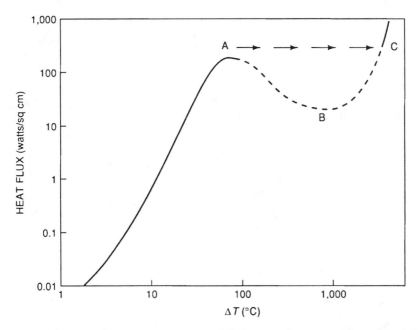

Figure 4.3A Heat flux versus temperature difference (surface temperature minus bulk water temperature) for a 1-millimeter diameter electrically heated wire in boiling water at 1 atmosphere.

film-boiling heat transfer as the wire temperature increases. However, this region must be unstable, in the sense that a steady state cannot be maintained.

Imagine that we are momentarily at a steady state in the region between A and B. Then, if the electrical heat input to the wire were momentarily to increase very slightly, the temperature of the wire would rise slightly. This rise would reduce the wire's ability to transfer heat to the water, and the wire temperature would continue to rise, to point B. By similar reasoning, if the electrical heat input were momentarily to decrease very slightly, the wire temperature would drop, and would continue to drop until point A was reached. Thus, there would be instability between A and B.

However, when we do the same sort of analysis in the region between B and C, we find the opposite result. We have stability. A momentary increase in heat input would cause a momentary rise in temperature, which in turn would cause an increase in heat flux to the water. The temperature would then drop back to the initial condition soon after the heat input rate came back to its initial condition. Similarly stable behavior would result from a momentary decrease in heat input.

Professor: Everything you say is correct. Why, then, does the wire temperature jump from point A to point C, when the power to the wire is increased slightly, if the region between B and C is stable?

Second Student: Because the region between B and C cannot transfer the heat to the water fast enough. That is not a stability problem.

As a matter of fact, if I were at point A, with electrical input of 300 watts per square centimeter, at a steady state, and I were to increase the power *momentarily* to 320 watts per square centimeter and then *immediately* reduce it to 50 watts per square centimeter, I would then reach a steady state between points B and C, assuming that I had a wire with a melting point higher than about 1,000 degrees C."

4.4 A CATALYTIC PARTICLE

Professor: What additional facts do you require to solve this problem?

First Student: Clearly, we are dealing here with a balance between the rate of heat generation by chemical reaction at the surface of the particle and the rate of heat loss from the particle to the surrounding gas. We need to have mathematical expressions for these rates, and then we can calculate the temperature at which these rates are balanced.

Professor: I'll provide that information, but only in general terms because we are only concerned with the principles here and not with a precise numerical answer.

The rate of chemical heat release, Q_c, increases more or less exponentially with the absolute temperature T_p, of the particle (Arrhenius's law):

$$Q_c = A \exp\left(-B/T_p\right) \qquad 4.1$$

where A and B are constants characteristic of the reaction. A is not strictly constant but depends on the partial pressures of the reactant and product gases near the surface.

As long as the particle temperature is low, the reaction rate will be low and the partial pressures near the surface will be essentially the same as in the bulk gas, so A is essentially constant. However, at higher temperatures the boundary layer around the particle will become seriously depleted in reactant gases, so additional reactant must then diffuse from the bulk of the gas to the surface through this boundary layer. This occurence can cause a substantial decrease in reactant concentrations at the surface and a corresponding decrease in the reaction rate. This effect is only important at high surface temperatures, when the reaction rate is fast but not very relevant, since the rate of the process will be controlled by the rate of diffusion of reactants to the surface (or the rate of diffusion of reaction products away from the surface).

Accordingly, the rate of heat generation versus particle temperature is as depicted in Figure 4.4A.

As for the rate of heat *loss*, Q_l, from the particle to the surrounding gas, let's assume that it is described by Newton's law of cooling:

$$Q_l = C(T_p - T_g) \qquad 4.2$$

where C is a constant that depends on the size of the particle and the thermal conductivity of the surrounding gas. (Of course, if the particle were very hot, there would also be radiative as well as conductive heat loss, but we will ignore that factor.)

With this information, how will you solve the problem?

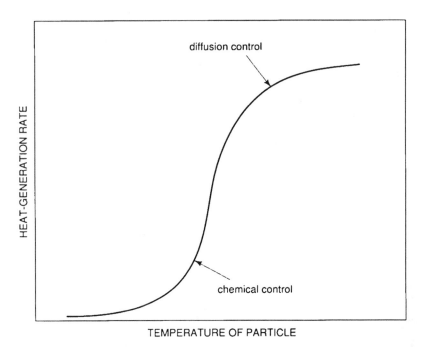

Figure 4.4A   Heat-generated rate versus temperature of a catalytic particle surrounded by reactive gas.

First Student: Very simply, in principle. I would equate Q_c in Equation 4.1 to Q_l in Equation 4.2 and solve for T_p. Unfortunately, when I do so, I find that I have a transcendental equation, which I cannot solve analytically for T_p because of the exponential term. However, given numerical values for A, B, C, and T_g, I can substitute these in the equation and solve it by trial and error, thus obtaining the steady-state value of T_p. This result would be valid as long as T_p came out to be a low enough value so that diffusion was unimportant.

Professor: Does anyone have a different approach?

Second Student: I would try a graphical solution. Notice that Equation 4.2 is the equation of a straight line. I would superimpose this heat-loss line on the heat-generation line of Figure 4.4A and determine the temperature at which they intersect.

Such a plot is shown in Figure 4.4B. But wait a minute! I see not one, but three intersections! What is going on?

First Student: Figure 4.4A includes the diffusion resistance effect at high temperatures, which causes the curve to decrease in slope and produces the additional intersections.

Second Student: I don't agree. You are right insofar as intersection 3 on Figure 4.4B is concerned, but intersections 1 and 2 would occur anyway, even if the change of slope due to the diffusion effect were not present. This fact is evident from looking at the figure.

Professor: First, you must realize that there is no mathematical requirement that every problem have a unique solution. Recall the simple equation $x^2 - 5x + 6 = 0$. Either $x = 2$ or $x = 3$ is a solution. Therefore, accept the demonstration of Figure 4.4B that three solutions are possible for that problem.

However, we want to investigate the stability of these three solutions.

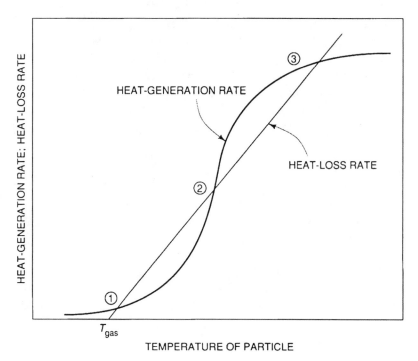

Figure 4.4B Balance of heat-generated rate and heat-loss rate for a catalytic particle.

The solution labeled 2 is clearly unstable because if the particle were at point 2, and a momentary, very small increase in particle temperature occurred, it would cause the rate of heat generation to become greater than the rate of heat loss; thus there would be no restoring tendency, and in fact, the temperature would continue to rise to point 3.

Similarly, a momentary, very small decrease in temperature would cause a drop all the way from point 2 to point 1 since throughout that region the rate of loss is greater than the rate of generation.

An investigation of points 1 and 3 will show that they are both stable.

Third Student: If points 1 and 3 are stable, how does the particle know whether to go to point 1 or point 3?

Professor: You recall that the original temperature of the particle was said to be T_{p0}. This temperature could be controlled by the experimenter, if this were a laboratory experiment. By inspection of Figure 4.4B, it should be clear that whenever T_{p0} is lower than the temperature at point 2, the particle will assume the temperature at point 1. And whenever T_{p0} is higher than the temperature at point 2, the particle will end up at point 3.

Fourth Student: Let's suppose that the gas temperature were to be increased from T_{g1} to T_{g2}. This change would shift the heat-loss curve but would not affect the heat-generation curve. (See Figure 4.4C.)

We note that we now get only one intersection instead of three, as long as T_g is greater than some critical value. Does this make sense, physically?

Professor: Absolutely. If the particle is stable at the temperature corresponding to point 1 and then the gas temperature is gradually increased, the particle temperature also increases moderately, until the loss curve is tangent to the generation curve. Then we are at a point of instability, and any further slight increase in T_g or in the particle temperature will cause a major temperature increase of the particle to point 3.

This result is analogous to the easily observable phenomenon of spontaneous ignition.

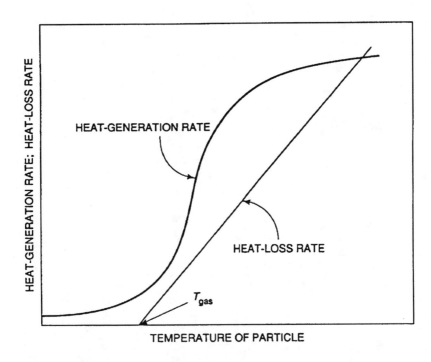

Figure 4.4C   Balance of heat-generation rate and heat-loss rate for a catalytic particle, with slightly higher ambient temperature than previous case.

CHAPTER 5

FINDING THE OPTIMUM

5.1 INTRODUCTION

We all recall from our study of calculus that we can find the maximum or minimum of a continuous function by differentiating the function and setting the first derivative equal to zero. We can determine if we are dealing with a maximum or a minimum by noting whether the second derivative is negative or positive. Of course, a function may have several maxima and/or minima. Also, a function may not be continuous.

Aside from the formal mathematical approach, there is some merit in trying to develop our intuition to be able to sense in what physical cases an optimum may exist. Also, it is useful, for those problems that are not easily reduced to precise mathematical formulation, to learn how to estimate the optimum.

Accordingly, some problems involving examples of optimization will be presented.

5.2 DRAINING THE TANK

An emergency arises and it becomes necessary to drain a hazardous liquid from a large steel tank as rapidly as possible. Unfortunately, the drain valve is corroded and cannot be opened. Also, because of the emergency, no electric power is available to operate an electric drill.

The only way to drain the tank is to crawl under it with a hand-operated drill and make a hole in the steel bottom. The liquid will then drain through the hole into an improvised trough.

Because of the hazardous nature of the liquid, I must leave the area immediately after drilling the first hole. Therefore, I can drill only one hole. What size hole should I drill?

If I drill a large hole through the steel with my hand-operated drill, the time to drill the hole, which will be proportional to the cross-sectional area of the hole, will be excessive. (A 1-sq-cm hole will require 10 minutes; a 5-sq-cm hole will require 50 minutes; etc.)

On the other hand, if I drill a small hole, the time needed for the liquid to drain will be excessive. The drainage time will be inversely proportional to the cross-sectional area of the hole. (A 1-sq-cm hole will result in a drainage time of 50 minutes; a 5-sq-cm hole will result in a drainage time of 10 minutes; etc.)

I don't have time for a detailed mathematical derivation, but I must make a decision quickly. I have 1 minute to think. What size hole do I drill?

READER: You have 2 minutes. Then turn to page 65.

5.3 INSULATING THE ELECTRIC CABLE

A bare copper electric cable surrounded by air, when carrying a very large current, will become quite hot because of ohmic heating. A man claims that if this cable is encased in plastic insulation, the copper will then operate at a significantly *lower* temperature while carrying the same current.

Can this statement be true? If so, is it true only under certain conditions? What would these conditions be?

READER: Can you provide an explanation of such an effect based on physical principles? Can you derive an equation defining the domain in which such an effect would exist? Then turn to page 68 for the solution.

5.4 MINIMIZING AUTOMOBILE COLLISIONS

A city has a square grid pattern of streets. The city manager consults a traffic "expert" concerning the optimum speed for vehicular traffic.

The "expert" reports that he has made a mathematical analysis of the problem, and the result shows that for minimization of collisions at intersections, the optimum speed is infinity. Since infinite speed is not achievable, he says that the vehicles should travel as fast as possible.

READER: Clearly, this analysis is wrong. But by what logic could the "expert" have reached such a conclusion? Also, what would be a better way of analyzing such a problem (other than a study of traffic statistics as a function of vehicular speed)? Then turn to page 73 for the solution.

5.5 HITTING A GOLF BALL

I want to hit a golf ball as far as possible. I reason that the greater the velocity of the club head when it makes contact with the ball, the further the ball will go. The mass of the ball is fixed at 40 grams. I want to know if there is an optimum mass for the head of the club, in order for me to get maximum range.

Unfortunately, my strength is finite, so the heavier the club the lower the velocity I can impart to it. If there is a roughly inverse proportionality between club mass and velocity, the momentum of the club head just before impact (mass x velocity) would appear to be nearly independent of the club mass.

Is there an optimum mass for the club head? If so, how would I calculate it?

READER: Try your hand at this one. Then turn to page 76.

SOLUTIONS TO CHAPTER 5
FINDING THE OPTIMUM

5.2 DRAINING THE TANK

Professor: Who has a quick answer?

First Student: From the data given, the time needed to drill a hole of area A is $10A$ minutes. The time to drain the tank is $50/A$ minutes.

My intuition tells me that if I select a drill size such that the time to drill is equal to the time to drain, neither time will be excessive and I will be either at the optimum or fairly close to it. Therefore, I equate the two times:

$$10A = 50/A$$

Solving for A, I see at once that A equals the square root of 5, or 2.236 square centimeters. Equating this area to $\pi d^2/4$, I find that d, the drill diameter, is 1.69 centimeters.

The time needed to drill this hole is 10 times 2.236, or 22.36 minutes. The time to drain the tank is the same, so the total time elapsed is 44.72 minutes, which I hope is near the optimum.

Professor: As a matter of fact, your solution is precisely correct. Let's derive the equation proving it. Let t_1 be the drilling time, t_2 the draining time, and t the total time. Then,

$$t = t_1 + t_2 = 10A + 50/A \qquad 5.1$$

Differentiation of t with respect to A, and setting the derivative equal to zero, as we were taught in calculus class, gives

$$dt/dA = 10 - 50/A^2 = 0 \qquad 5.2$$

Solving for A, we find

$$A = \sqrt{[50/10]} = 2.236 \text{ sq cm} \qquad\qquad 5.3$$

which is exactly the answer the first student obtained intuitively.

Figure 5.2A shows this solution graphically, and it is clear that the minimum in the upper curve, corresponding to zero slope or zero first derivative, is the desired optimum.

Second Student: I have a question. Was the first student simply lucky in getting the correct answer without the use of calculus? Or is there some underlying principle that, if we knew it, could be used to solve such problems without the use of calculus?

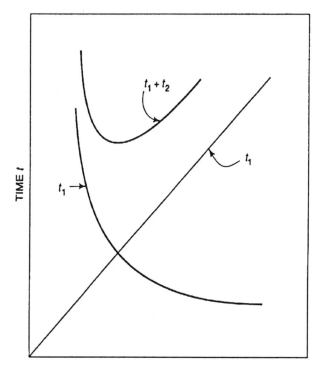

Figure 5.2A   Drilling time, t_1, and draining time, t_2, versus area of hole in tank to be drained.

Professor: That is a good question. Let me answer it by presenting a derivation of a more general version of the same problem.

Suppose that we are seeking the minimum value of t $(= t_1 + t_2)$, where t_1 and t_2 are given by

$$t_1 = k_1 A^m \text{ and } t_2 = k_2/A^n \qquad 5.4$$

k_1, k_2, m, and n being positive constants. (In the former, less general case, m and n were each equal to 1.) Then,

$$t = t_1 + t_2 = k_1 A^m + k_2/A^n \qquad 5.5$$

Differentiating and setting the derivative equal to zero, we get

$$dt/dA = mk_1 A^{m-1} - nk_2 A^{-n-1} = 0 \qquad 5.6$$

Solving Equation 5.6 for the optimum value of A, we find

$$A_{\text{optimum}} = [(nk_2)/mk_1)]^{1/(m+n)} \qquad 5.7$$

Suppose we had ignored calculus and sought to calculate, in this general case, the optimum area simply by equating t_1 to t_2. Such an approach would give us

$$A^* = [k_1/k_2]^{1/(m+n)} \qquad 5.8$$

Now let's compare A_{optimum} with A^* by taking the ratio of Equation 5.7 to Equation 5.8:

$$A_{\text{optimum}}/A^* = [n/m]^{1/(m+n)} \qquad 5.9$$

Notice that when m and n are each unity, as was the case in the original problem, Equation 5.9 tells us that A_{optimum} is the same as A^*. However, the equation also tells us that even if m and n are not unity but if m is the same as n, the ratio n/m is unity and therefore A_{optimum}/A^* is unity.

But if m is *not* equal to n, Equation 5.9 requires A_{optimum} to be different from A^*. For example, if $m = 0.5$ and $n = 1.5$, then $A_{optimum}/A^*$ comes out to be $\sqrt{3}$.

Second Student: On the basis of your presentation, I must conclude that the first student was lucky in his approach, insofar as getting the precisely correct answer; but nonetheless that approach is useful in getting a reasonable approximation to the answer when time does not permit a thorough analysis.

5.3 INSULATING THE ELECTRIC CABLE

Professor: Does anyone believe the man's claim?

First Student: We all know that the thermal conductivity of plastic insulation is several orders of magnitude smaller than that of copper. It is evident that as we surround the copper with insulation, we provide a barrier to the escape of heat. Hence the copper must get hotter, not cooler, to get rid of the ohmic heat. I don't believe that the man's claim can be true.

Second Student: I don't think you would have presented this problem to us unless there is more to it than the first student says. Let's analyze it carefully.

The bare cable loses heat by convection to the surrounding air. There is a boundary layer or film of stagnant air around the cable that acts as a resistance to escape of the heat. The greater the wind velocity, the thinner is this film, so less resistance to heat transfer would be offered if a wind blew across the cable, and its temperature would be lower. Also, of course, the resistance to heat transfer would be inversely proportional to the exposed surface area of the cable, other factors being the same.

Now, if we add a fairly thick layer of insulation around the cable, we will have two resistances to heat transfer, in series. The heat has to flow through the insulation, which has a fairly low thermal conductivity, and also through the air film, which has a *very* low thermal conductivity. However, since the exposed surface area of the insulated cable can be much larger than that of the bare cable, maybe two or three times larger, it follows that the resistance due to the air film is less for the insulated cable than for the bare cable. We need to know how this decrease in resistance compares with the additional resistance caused by poor conductivity of the insulation.

Until we do a calculation, we must leave the door open to the possibility that some optimum thickness of insulation will give a minimum temperature for the interior of the cable that is lower than that of the bare cable.

Third Student: That makes sense to me, but I want to point out that if the cable gets quite hot, it will lose heat by radiation as well as by convection to the air, and this fact should be taken into account.

Second Student: Very true. But radiative heat transfer as well as convective heat transfer from the outer surface will each increase as the surface area increases because of a larger diameter, so the qualitative argument I presented would not be affected if radiation from the outer surface were dominant.

Professor: The points you make are good. But we need a quantitative treatment to find out under what conditions, if any, the temperature inside the cable will drop as insulation is added. I shall present the relevant equations.

Let W be the ohmic heat generation (watts) in a cable of length L. Let r_1 and r_2 be the copper radius and the outer radius of the insulation, respectively. Let T_1, T_2, and T_3 be the temperatures of the copper, the outer surface of the insulation, and the surrounding air, respectively. Let k be the thermal conductivity of the insulation, and let h be the heat transfer coefficient for heat flow from the outer surface to the air, defined by the equation

$$W = hA(T_2 - T_3) = h(2\pi r_2 L)(T_2 - T_3) \qquad 5.10$$

Of these variables, we consider W, L, r_1, T_3, k, and h to be known constants. (Actually, h would vary slightly with surface temperature and with r_2, but we ignore this factor. Also, we do not explicitly consider radiative heat loss, although it could, as an approximation, be lumped into h.) We consider T_1, T_2, and r_2 to be unknown variables. We are interested in the relationship between T_1 and r_2.

In the steady state, the heat that flows through the insulation also flows through the air film. We express this by the equation

$$W = (T_1 - T_2)/R_{\text{ins}} = (T_2 - T_3)/R_{\text{film}} \qquad 5.11$$

In this equation, R_{ins} is the resistance of the insulation to heat transfer, which since we are dealing with a cylindrical geometry, is given by

$$R_{\text{ins}} = ln(r_2/r_1)/(2\pi k L) \qquad 5.12$$

as derived in heat transfer textbooks; R_{film} is the resistance of the air film to heat transfer, which as is obvious from Equation 5.10, is given by

$$R_{film} = 1/(2\pi r_2 Lh) \qquad\qquad 5.13$$

Substituting Equations 5.12 and 5.13 into Equation 5.11 and solving for T_2, we obtain

$$T_2 = [T_1 + T_3(r_2 h/k)ln(r_2/r_1)]/[1 + (r_2 h/k)ln(r_2/r_1)] \qquad 5.14$$

We can eliminate T_2 and obtain an equation for T_1 in terms of the other variable, r_2, by substituting Equation 5.14 into Equation 5.10. The resulting equation, after rearranging, becomes

$$T_1 = T_3 + [(k/r_2 h) + ln(r_2/r_1)][W/(2\pi kL)] \qquad\qquad 5.15$$

Let's study this equation. The terms in the second pair of square brackets are all constants, as is T_3; the two terms in the first pair of square brackets each contains r_2, in the denominator of the first term and in the numerator of the second term. Thus, it seems possible that an increase of r_2 could cause either an increase or a decrease of T_1.

Let's consider some limiting cases. In the limit of no insulation, $r_2/r_1 = 1$, its logarithm equals zero, and the equation reduces to

$$T_1 = T_3 + W/(2\pi r_1 Lh) \qquad\qquad 5.16$$

In the other limit of infinitely thick insulation, Equation 5.15 reduces to T_1 = infinity since the logarithm of r_2 is infinite.

We know, then, that a plot of T_1 versus r_2 would start at r_1 with the T_1 value given by Equation 5.16 and would go to very large values of T_1 at sufficiently large values of r_2, but we don't yet know if T_1 would pass through a minimum on the way. We determine the answer by differentiating Equation 5.15, obtaining

$$dT_1/dr_2 = (1/r_2)[1 - (k/hr_2)] [W/(2\pi kL)] \qquad\qquad 5.17$$

Finally, we have the equation that gives us the whole story. Notice that for a given value of k/h, the term k/hr_2 has its largest value when there is no insulation, when $r_2 = r_1$. Accordingly, if k/hr_1 is less than unity, all values of k/hr_2 will be less than unity, and Equation 5.17 requires that dT_1/dr_2 be always a positive number.

On the other hand, if k/hr_1 is appreciably greater than unity, it follows from Equation 5.17 that dt_1/dr_2 must be negative as long as r_2 is only slightly greater than r_1; it then changes from negative to positive for large values of r_2, the transition from negative to positive slope occurring when $r_2 = k/h$. (This statement would be proven formally by setting the right-hand side of Equation 5.17 equal to zero and solving for r_2.)

Figure 5.3A shows plots of interior cable temperature, T_1, versus outer radius of insulation, r_2 for k/hr_1 to equal 0.5 and 2.0. (Reasonable values have been used for the other parameters.)

Our final finding is that whenever k/hr_1 is greater than unity, the temper-

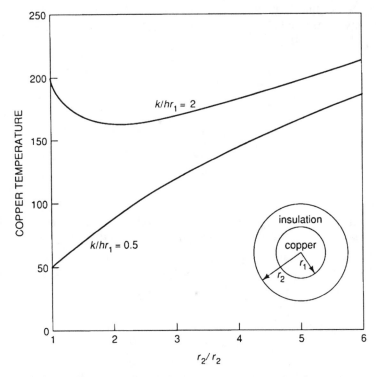

Figure 5.3A   Variation of copper temperature with insulation thickness, for two values of k/hr_1.

ature inside the cable will be at a minimum value (lower than for the uninsulated cable) when the insulation thickness is such that $r_2 = k/h$. The internal temperature at that condition is easily obtained from Equation 5.15:

$$(T_1)_{min} = T_3 + [1 + ln(k/hr_1)][W/(2\pi kL)]$$

5.18

By taking a ratio based on Equation 5.18, for the optimally insulated wire, and Equation 5.16, for the bare wire, we can see how much lower the interior temperature of the optimally insulated wire is:

$$[(T_1)_{min} - T_3]/[(T_1)_{bare} - T_3] = [1 + ln(k/hr_1)]/[k/hr_1]$$

5.19

Remember that Equation 5.19 is only valid for k/hr_1 equal to or greater than unity. From this equation, we can calculate the temperature ratio for various values of k/hr_1:

k/hr_1	$[(T_1)_{min} - T_3]/[(T_1)_{bare} - T_3]$	$(r_2/r_1)_{opt}$
1	1	1
2	0.847	2
3	0.699	3
4	0.597	4

The values in the last column follow from the relation $(r_2)_{opt} = k/h$. Thus, we see that if we had an insulated cable with k, h, and r_1 values such that $k/hr_1 = 4$, and we provided sufficient insulation so that the outer radius was four times the copper radius, the temperature rise for a given power dissipation rate would be only 59.7 percent as great as for a bare conductor.

On the other hand, if we had an insulated cable such that k/hr_1 was less than 1, the copper temperature with insulation would always be higher than for the bare wire, regardless of the insulation's thickness.

For your information, the dimensionless ratio hr/k, which is so important in this problem, is known as the Biot number.

5.4 MINIMIZING AUTOMOBILE COLLISIONS

Professor: Can anyone reproduce the logic of the "expert"?

First Student: Let's assume that there are n motorists in the city, and each of them travels k kilometers each day. If they all drive at a speed of s kilometers per hour, each motorist is driving for k/s hours each day. In total, the n motorists drive for nk/s hours each day.

Let's assume that virtually no one drives between midnight and 6:00 A.M., so the daily driving occurs at random times for each motorist over an 18-hour period each day.

Thus, the average number of motorists driving at any instant in the 18-hour period must be $nk/(18s)$. The average density of cars moving on the streets is seen to be inversely proportional to s, the average speed.

This fact is understandable because if everyone drove a given distance each day, at random times, at a given speed, and then this speed were doubled, each motorist would be driving for only half as long. Thus, half as many cars would be in motion at any instant.

With fewer cars in motion at any instant, we might expect a lower probability of collisions. To carry this reasoning to an extreme, let's assume some numerical values:

$$\text{Let } n = 36{,}000 \text{ motorists}$$

$$k = 50 \text{ kilometers}$$

$$s = 100{,}000 \text{ kilometers per hour}$$

With these numbers, the number of motorists driving at any instant, anywhere in the city, is equal to $nk/(18s) = 1$. If, on the average, only one motorist is driving at any instant, clearly he or she doesn't have to worry about collisions.

Professor: Although none of us would take the answer seriously, the logic seems to be self-consistent. Who is able to carry it a step further and develop a formula showing how the frequency of collisions varies with the speed?

Second Student: Consider a given motorist during a small time interval. His or her probability of having a collision is proportional to the number of targets (other motorists) moving on the streets in that time interval. (Let's ignore collisions with parked cars.)

But during that interval, $nk/(18s)$ motorists are in motion. The probability that a collision will occur during the interval is proportional to the *product* of the number of motorists in motion and the number of targets for each of these motorists, which in turn is equal to the number of motorists in motion (minus 1, which we ignore). Therefore, the probability of a collision during this small time interval is proportional to the *square* of the number of motorists in motion:

$$[(nk)/(18s)]^2$$

However, the probability of collisions also depends on another factor. The faster a vehicle is going, the greater is the chance that it will pass through a space occupied by another vehicle during the small time interval. Hence, because of this factor, the probability of a collision is proportional to s. When this factor is combined with the previous factor by multiplying probabilities to obtain the overall probability, the result is

$$(nk/18)^2/s$$

Therefore, since n and k are constants, the probability of a collision at any instant is seen to be inversely proportional to the speed, s. That is what you asked for, isn't it?

Professor: Very good. Now let's ask what is wrong with this analysis.

First Student: Although an increase in the speed limit would decrease the number of collisions per day, it would obviously increase the violence of each collision that took place, and hence would increase the fatalities.

Professor: It is certainly true that high-speed collisions will result in more fatalities. But the city manager wanted to decrease the number of collisions and didn't say anything about fatalities. Do you really think that the number of collisions would decrease if the speed limit were increased?

Third Student: I wish to analyze the problem a different way. I assume that I am concerned only with intersections in the city that are not protected by traffic lights or stop signs. Assume that two vehicles, one moving north and one moving east, are approaching an intersection in such a way that a collision would occur unless one takes evasive action, for instance by stopping.

Now, let t_1 be the interval between the time when one driver sees the other vehicle and becomes aware of the need for action (the detection time) and the time at which the collision would occur if no effective action were taken.

Let t_2 be the time necessary to stop one of the vehicles, once the driver has detected the emergency. Obviously, if t_1 is greater than t_2, there will be no collision, and if the reverse is true, there will be a collision.

How do t_1 and t_2 depend on the speed of the vehicles? Obviously, the greater the speed, the greater the braking time and the greater the value of t_2. Since t_2 is the sum of the driver's physical reaction time and the vehicle's response time to brake application, and only the latter depends on speed, we would not expect a direct proportionality between t_2 and speed, but we would expect a positive dependence.

As for t_1, consider Figure 5.4A. Because of the building on the southwest corner of the intersection, the driver of the northbound vehicle cannot see the other vehicle until his or her car is a distance, x, from the potential collision point. Thus, the available time, t_1, for collision avoidance is equal to x/s. Clearly, t_1 is inversely proportional to s, whereas as we have seen, t_2 increases when s increases. Figure 5.4B shows these relationships. According to that figure, no collision will occur as long as the speed is less than the critical value where the curves cross one another. This statement, of course, assumes that both drivers are alert and both vehicles are in good mechanical condition.

Professor: Excellent. The difference between your analysis and the previous one is the implicit assumption of the first analysis that the cars are moving randomly and mindlessly. It so happens that the kinetic theory of gases predicts a molecular collision frequency that is proportional to the square of the number of molecules present per unit volume and to the molecular speed, in agreement with the first analysis. However, molecules approaching one another cannot apply the brakes.

As a final comment on this problem, a more sophisticated analysis than that of the third student would take into account the *distributions* of driver reaction times, braking times, corner geometries, etc., each of which could be represented as bell-shaped curves.

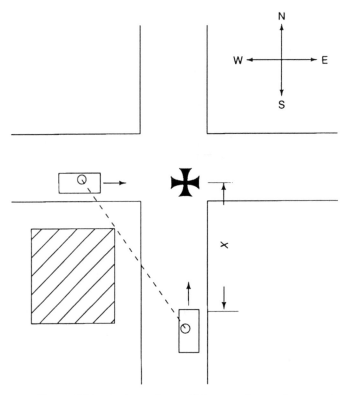

Figure 5.4A   An impending collision at an intersection.

5.5 HITTING A GOLF BALL

Professor: I'll give you 20 minutes to work on this problem.

(20 minutes later)

First Student: I have made an analysis. Let's suppose that a club head of mass m_c, moving at velocity v_1, contacts a ball of mass m_b. These quantities are assumed to be known. The resulting velocity of the ball, v_b, and the velocity of the club head after impact v_2, (which must be less than v_1) are unknowns. To determine these two unknowns we need two equations. We know that both momentum and energy are conserved, which gives us the two relevant equations:

$$\text{Momentum: }\; m_c (v_1 - v_2) = m_b v_b \qquad\qquad 5.20$$

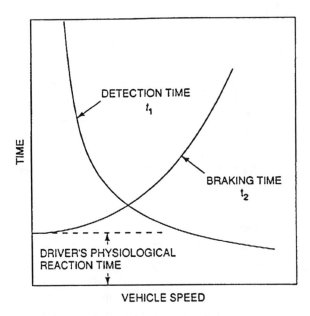

Figure 5.4B Variation of detection time and braking time with vehicle speed.

$$\text{Energy: } m_c \, (v_1{}^2 - v_2{}^2)/2 = m_b v_b{}^2/2 \qquad\qquad 5.21$$

When we solve these equations simultaneously, eliminating v_2, in which we are not interested, we obtain a value for v_b:

$$v_b = 2(v_1 m_c)/(m_c + m_b) = 2(\text{momentum of club head})/(m_c + m_b) \quad 5.22$$

Assuming that the momentum of the club is fixed for a golfer of given strength, we see from Equation 5.22 that v_b, the velocity of the ball, increases as m_c decreases and approaches its maximum as m_c approaches zero.

Hence I conclude that the lighter the club head the further the ball will go.

Professor: Did anyone reach a different conclusion?

Second Student: We should also solve the foregoing equations, 5.20 and 5.21 for v_2, the velocity of the club head after impact. When we do so, we obtain

$$v_2 = v_1[(m_c - m_b)/(m_c + m_b)]$$

5.23

Equation 5.23 tells us that v_2 will be zero if $m_c = m_b$, positive if m_c is greater than m_b, and negative if m_c is less than m_b. If we watch a professional golfer hit a ball a long distance, we see that v_2 is always positive (follow-through), so we must conclude that in practice, m_c is always greater than m_b. This result is at variance with the previous finding that m_c should approach zero.

Does the result mean that currently used golf clubs are not properly designed?

Professor: If you study a golf club designed to hit a ball a long distance (a driver), you will find that the club weighs roughly ten times as much as a golf ball. Further, you will find that about 45 percent of the weight of the club is in the head, and about 55 percent is in the shaft, which is about 1.1 meters long.

It seems logical that an ideal club should have a weightless shaft, with all the weight in the club head, so that maximum velocity of the club head can be achieved. However, the shaft must be strong enough to withstand the impact and also rigid enough so that it can be controlled by the golfer, with the result that the ball is hit squarely. There is another factor I will mention in a moment. Because of these requirements, the shaft must weigh at least about 200 grams.

The club head cannot be too light relative to the weight of the club shaft because the club can be swung more accurately if a substantial part of the weight is concentrated at the far end. In other words, the club is better balanced if this rule is followed.

But this is still not the whole story. Imagine that a golfer *without* a club goes through the motion of swinging a club as fast as possible. The speed with which the hands can move in this case is clearly limited by the inertia of the hands, the arms, and the pivoting portion of the body. *With* the club, the speed will be somewhat slower, but clearly the mass of the golfer is a factor as well as the mass of the club.

This brings us to a conclusion: For a golfer of given strength and mass, with a club shaft of given mass, the achievable velocity of the club head is only moderately dependent on the club head mass, as long as it is small compared with the shaft mass and body mass. Furthermore, if the shaft is rigid, some of the momentum of the shaft and the twisting body of the golfer is transmitted through the shaft to the ball, as well as momentum from the club head itself. (This is another reason why the shaft cannot be weightless.)

Accordingly, an extremely heavy club head cannot be swung rapidly, and an extremely light club head cannot be swung accurately. In between, determination of the best weight for the club head must be based on testing by golfers since human physiology is involved. Precise calculations of optimum mass are not possible.

Third Student: In the foregoing analysis, we have assumed that the range of the golf ball increases with increasing initial velocity, v_b. Is this a linear relationship? Another kind of relationship? Also, wouldn't the range depend on the angle between the initial trajectory and the ground?

Professor: This is something we can analyze precisely, in a very simple way, if we neglect the atmospheric drag.

READER: Do you want to make this analysis? Then turn to page 80.

Professor: Figure 5.5A shows a trajectory, with drag neglected. The golf ball has an initial velocity v, resolvable into x and y components, v_x and v_y, and initially moves at an angle, θ, with the horizontal. The desired range is x_1, and the maximum height achieved is y_1.

If a ball were to be projected vertically upward at initial velocity, v_y, and drag is neglected, it would reach its maximum height, y_1 in a time t_0, and since it is being decelerated by gravity at the rate of g meters per second each second, the following relationship must hold:

$$v_y - gt_0 = 0 \ \text{ or } \ t_0 = v_y/g \qquad\qquad 5.24$$

Similarly, if the same ball were dropped from height y_1 and fell to the ground, the time required would also be $t_0 = v_y/g$

Applying this formula to Figure 5.5A and noting that the horizontal and vertical components of velocity may be treated separately, we see that the time of flight of the golf ball, neglecting drag, must be $t_0 + t_0$, or $2v_y/g$.

In the absence of drag, the *horizontal* velocity component, v_x, of the ball is constant. During the time $2v_y/g$, which is the time of flight, the ball will move a horizontal distance, x_1 given by

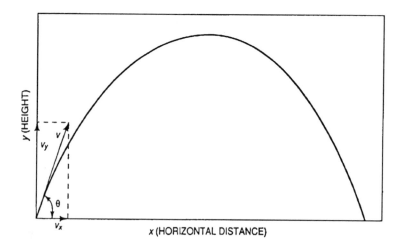

Figure 5.5A The trajectory of a ball with no drag (v is initial velocity and θ is initial angle).

$$x_1 = \text{distance} = \text{velocity x time} = (v_x)(2v_y/g) = 2v_xv_y/g \qquad 5.25$$

But

$$v_x = v(\cos \theta) \text{ and } v_y = v(\sin \theta) \qquad 5.26$$

Therefore

$$x_1 = 2v^2(\sin \theta)(\cos \theta)/g \qquad 5.27$$

Equation 5.27, plotted in Figure 5.5B, shows how the range depends on θ as well as on v. We can determine the optimum value of θ that will maximize x_1, for a given value of v, by differentiating Equation 5.27 and setting the derivative equal to zero:

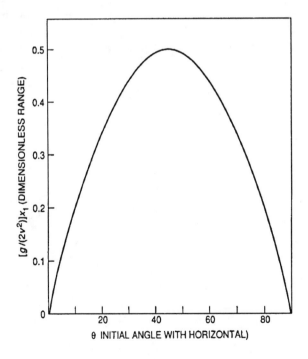

Figure 5.5B   The range of a ball as a function of initial angle, θ, with no drag.

$$dx_1/d\theta = (2v^2/g)(\cos^2 \theta - \sin^2 \theta) = 0 \qquad\qquad 5.28$$

Obviously, $dx_1/d\theta = 0$ only when $\cos^2 = \sin^2\theta$. This occurs when $\theta = 45$ degrees. We can prove that the value of x_1 corresponding to $\theta = 45$ degrees is a maximum rather than a minimum by obtaining the second derivative, $d^2x_1/d\theta^2$, and observing that it is negative:

$$d^2x_1/d\theta^2 = -8v^2[(\sin \theta)(\cos \theta)]/g \qquad\qquad 5.29$$

When the golf ball is struck so that it acquires an initial velocity, v, and an initial angle of 45 degrees with the horizontal, Equation 5.27 reduces to

$$(x_1)_{max} = v^2/g \qquad\qquad 5.30$$

since $\sin 45° = \cos 45° = \sqrt{2}/2$

Thus, when the angle is optimum, the range is very simply related to the square of the velocity, with no drag. When drag is considered, a much more complex calculation is required to determine the dependence of range on v and θ, which requires numerical rather than analytical solution of the equation of motion.

CHAPTER 6

SOME SCIENTIFIC CURIOSITIES

6.1 SURFACE-TO-VOLUME RATIO

It is well known that the surface-to-volume (S/V) ratio of a solid object depends on its shape. Of all possible shapes, a sphere has the lowest S/V ratio.

Observation shows that an isolated soap bubble, which has surface tension, assumes a spherical shape, except for distortions possibly caused by aerodynamic drag or by gravitation. Similarly, a small drop of water, under ideal conditions, is round.

Assume that a sphere of diameter d is placed in a cube of dimension d. Calculate the S/V ratios for the sphere and for the cube, in terms of d. How much smaller is the S/V ratio for the sphere? Is your answer reasonable?

READER: Please turn to page 87 after solving the problem.

6.2 THE LEANING TOWER OF PISA

Let's imagine that Galileo had a billiard ball and a tennis ball of identical size, so that the aerodynamic resistance acting on either ball moving at the same velocity would be identical. However, the billiard ball was ten times as heavy as the tennis ball.

Suppose that Galileo simultaneously dropped these two balls from the top of the leaning tower of Pisa, on a windless day. Would they reach the ground at the same time?

READER: What do you think? Then, turn to page 88.

6.3 ABSOLUTE TEMPERATURE

Most of us have learned that there are two primary reasons why the absolute, or Kelvin, scale of temperature was introduced:

> 1. The volume occupied by a gas at constant pressure is directly proportional to its absolute temperature. (This proportionality holds as long as the pressure is well below the critical pressure of the gas and/or the temperature is well above its critical temperature.)

> 2. Zero on the absolute temperature scale is the lowest possible temperature.

However, in addition, there are many physical and chemical phenomena with temperature dependencies that can be described more simply on the absolute temperature scale, T (Kelvin), than on the Celsius scale, t. (Obviously, since $T = t + 273.1$, any mathematical dependence on T can also be expressed in terms of t, although in a somewhat more complicated way.)

Can you name at least a dozen phenomena, in addition to the two mentioned, that depend more directly on T (K) than on t (C)?

READER: How many can you name? Then, turn to page 90.

6.4 A MATTER OF GRAVITY

The gravitational attractive force, F, between two masses, M_1 and M_2, separated by a distance, X, is given by

$$F = 6.67 \times 10^{-11} M_1 M_2 / X^2 \text{ newtons}$$

where the masses are expressed in kilograms and the distance in meters.

In the case of the attraction between the earth and an apple above the earth's surface, Newton proved that the distance X should be taken as the distance from the apple to the center of the earth, which is the "center of mass" of the earth.

Consider a large, concrete, hollow sphere having a mass of 10,000 kilograms. Consider the gravitational interaction between this hollow sphere and a small object having a mass of 0.05 kilograms, located as shown in Figure 6.4A. The distance, X, between the small object and the center of mass of the concrete shell is 0.1 meters.

Calculate the attractive force between these two bodies. Ignore the earth's gravitational field.

READER: You have 2 minutes to solve this problem. Then turn to page 91.

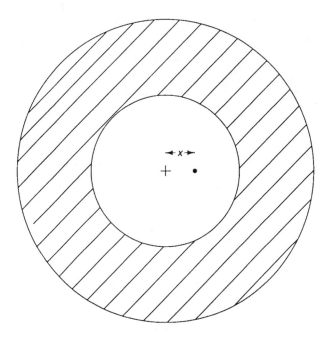

Figure 6.4A   A problem in gravitational attraction.

6.5 LETTING AIR INTO AN EMPTY TANK

I have used a vacuum pump to remove all the air from a spherical tank with a volume of 1,000 cubic meters. Outside the tank is air at 1 atmosphere. Both the air and the tank are at 20 degrees C.

I disconnect the vacuum pump and open a valve that allows the outside air to flow rapidly into the tank until the internal pressure reaches 1 atmosphere. This step requires 30 seconds.

Next I measure the temperature of the air at the center of the tank. What is this temperature?

READER: Unless you have studied thermodynamics, you have little chance of solving this problem. When ready, turn to page 95.

6.6 THE TWO CAPACITORS

Using direct current, I apply a voltage, V_0, to a capacitor of capacitance C, and it acquires a charge, Q_0. I then remove the charging source and I connect both sides of the capacitor to the two sides of another capacitor, which is identical to the first capacitor but is uncharged. I use low-resistance cables when making these connections.

After a short time, I find that the charges have equalized between the two capacitors, and each has a charge equal to $Q_0/2$. Furthermore, I find that the voltage across the terminals of each of the capacitors is equal to $V_0/2$.

I know that the energy stored in a capacitor is $CV^2/2$. Accordingly, I calculate that the energy that was originally in the first capacitor was $CV_0^2/2$, and the energy stored in the two capacitors after the charge equalization is

$$2[C(V_0/2)^2/2] = CV_0^2/4$$

Unless my calculation is wrong, half the energy has disappeared.

READER: Is there a trick here? If not, where did the energy go? After thinking about it, turn to page 98 for the solution.

SOLUTIONS TO CHAPTER 6
SOME SCIENTIFIC CURIOSITIES

6.1 SURFACE-TO-VOLUME RATIO

Professor: This problem is really extremely simple. Who has a solution?

First Student: The S/V ratio of a sphere is $\pi d^2/(\pi d^3/6) = 6/d$. The S/V ratio for a cube is $6d^2/d^3 = 6/d$. Obviously, the S/V ratio is the same for a sphere and a cube.

Professor: But how can you reconcile your answer with the generally accepted, and indeed obvious, fact that a sphere has a lower S/V ratio than a cube?

First Student: I can't.

Second Student: I can explain it. Suppose I made the following statement. If I take *a given volume* of clay, or other material, and mold it into a sphere, its S/V ratio will be lower than if I were to mold it into a cube. Doesn't that sound reasonable?

In the stated problem, the volume of the sphere was $\pi d^3/6$. The volume of the cube was obviously larger. If we wanted to construct a cube of the same volume as the sphere, its linear dimension would be the cube root of its volume, or $(\pi/6)^{1/3} d$. It is easy to calculate that the S/V ratio of such a cube would be 1.2407 x $(6/d)$, compared with $(6/d)$ for a sphere.

Thus, the S/V ratio of an equivalent cube is 24 percent greater than that of a sphere.

Professor: And what do we learn from this problem?

Second Student: Two things. First, make sure that a problem is stated precisely and logically before trying to solve it. And second, if your first answer does not agree with your common sense, take a very hard look at the problem and at your solution.

6.2 THE LEANING TOWER OF PISA

Professor: You have all studied physics. What is the answer?

First Student: We are certain that if the two balls were dropped in a vacuum, they would each increase speed by 9.8 meters per second each second and would reach the ground together. However, the presence of the atmosphere will produce an aerodynamic drag, which will reduce the rate of acceleration.

In this case, since the tennis ball is exactly the same size as the billiard ball and will experience exactly the same drag force at each velocity, I would expect that the two balls would fall together through the air.

Professor: Any other views?

Second Student: I must disagree. The billiard ball will reach the ground sooner. What produces the acceleration is the *net* force acting on each ball, which is the difference between the gravitational force and the drag force. The drag force is the same for the two balls at any given velocity, but the gravitational force on the billiard ball is ten times as great as that on the tennis ball; hence, as the velocity increases with increasing drag force, the drag force becomes significant relative to the constant gravitational force much sooner for the tennis ball and reduces its acceleration earlier than is the case for the billiard ball.

Third Student: I reach the same conclusion by slightly different reasoning. We know that either ball, if dropped from a sufficient height, will reach a terminal velocity when the drag force has increased to the point where it is equal to the gravitational force.

Obviously, the terminal velocity of the billiard ball will be much greater than that of the tennis ball. Since the drag is approximately proportional to the square of the velocity, it is easy to show that the terminal velocity of the billiard ball will be the square root of ten times as great as that of the tennis ball.

Clearly, then, once the two balls are released, they must follow different velocity histories in order to end up at different terminal velocities."

Professor: To quantify our understanding, let's look at the equation of motion. In a vacuum, according to Newton's second law of motion,

$$\text{force acting} = mg = m(dv/dt) \qquad\qquad 6.1$$

where m is the mass and v is the velocity of a ball, g is the local acceleration due to gravity (= 9.8m per sec per sec), and t is time. A glance at Equation 6.1 reveals that the m's cancel one another, and dv/dt (the acceleration) = g.

Now the same equation, for motion through air instead of a vacuum, becomes

$$\text{Force acting} = mg - kv^2 = m(dv/dt) \qquad 6.2$$

the aerodynamic drag being assumed proportional to v^2, or equal to kv^2, where k is an appropriate constant, the value of which depends on the diameter of the ball and the density of the air.

Obviously, m does not cancel out of Equation 6.2, so when the equation is solved, a functional relationship among v, m, and t will result. The terminal velocity is easily seen to be

$$v_{\text{terminal}} = \sqrt{m(g/k)} \qquad 6.3$$

The solution of Equation 6.2 is illustrated in Figure 6.2A.

Fourth Student: I have a question. Suppose that Galileo had dropped a billiard ball and a *golf* ball simultaneously. The weight of the golf ball might be 1/5 the weight of the billiard ball, but since the golf ball is smaller,

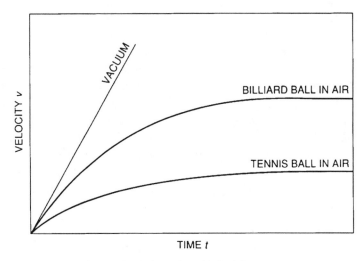

Figure 6.2A   Velocity histories of balls falling in vacuum and in air.

its aerodynamic drag might also be 1/5 that of the billiard ball, at a given velocity. Wouldn't such a pair of balls reach the ground simultaneously?

Professor: Let's rearrange Equation 6.2 by dividing through by m, obtaining

$$dv/dt = g - (k/m)v^2 \qquad\qquad 6.4$$

Your assumed pair of balls is equivalent to adjusting the ratio k/m to be identical for the golf ball and the billiard ball. It follows, of course, from Equation 6.4 that the velocity history of both balls would have to be the same, and they would reach the ground together.

6.3 ABSOLUTE TEMPERATURE

Professor: I ask each student to give us three such phenomena.

First Student: First, so-called black body radiation intensity varies directly with the fourth power of the absolute temperature.

Second, the velocity of a sound wave propagating through a gas varies directly with the square root of the absolute temperature.

Third, the maximum possible efficiency of a heat engine with heat intake at temperature T_i and rejection of unconverted heat at temperature T_r is given by $[1 - (T_r/T_i)]$, where the temperatures must be on the absolute scale.

Second Student: I shall draw on my background in chemistry.

First, van't Hoff's equation shows how the equilibrium constant for any chemical reaction varies with temperature. It states that the negative logarithm of the equilibrium constant is inversely proportional to the absolute temperature. (The equilibrium constant is the ratio of partial pressures of gases in equilibrium.)

Second, Arrhenius's equation for the dependence of the rate constant of any chemical reaction states that the negative logarithm of the rate constant is inversely proportional to the absolute temperature.

Third, Trouton's rule states that the latent heat of vaporizaton per mole of any nonpolar liquid is about 88 T_B joules, where T_B is the boiling point, expressed in absolute temperature, of the liquid at 1 atmosphere.

Third Student: I have recently been studying the kinetic theory of gases. The following relations are valid for any monatomic gas, such as helium, neon, or argon, and are approximately true for diatomic gases:

First, the viscosity coefficient is proportional to the square root of absolute temperature.

Second, the thermal conductivity coefficient is proportional to the square root of absolute temperature.

Third, the diffusion coefficient is proportional to the 3/2 power of absolute temperature.

Fourth Student: First, the translational energy of a gas molecule at absolute temperature, T, is equal to $3kT/2$, where k is Boltzmann's constant.

Second, the product of the critical pressure and the critical volume per mole of any nonpolar substance is very nearly equal to $3RT_c$, where R is the gas constant and T_c is the critical temperature (absolute units) of the substance.

Third, the osmotic pressure of a solution (e.g., sugar or salt in water) is proportional to the absolute temperature.

Professor: Well, that makes a dozen. I'll throw in one more. According to Curie's law, the paramagnetic susceptibility of substances is inversely proportional to absolute temperature.

Fifth Student: I have a question. The Celsius scale of temperature is purely arbitrary, being based on the freezing point and boiling point of water at sea level. On the other hand, the Kelvin scale of temperature is obviously based on something very fundamental since we have just listed 15 natural laws based on it. Why do we bother with the Celsius scale at all? We only need one temperature scale. Clearly, it should be the Kelvin scale.

Professor: The Celsius scale, also known as the Centigrade scale, was introduced early in the eighteenth century by the Swedish scientist Anders Celsius, who died in 1744. The 15 laws we have enumerated were not known at that time. In 1787, J. A. C. Charles discovered the first of these laws, the dependence of the volume of a gas on temperature. Other laws were discovered in the nineteenth and early twentieth centuries. However, the Celsius scale had become firmly established by that time. Because of inertia, it has not been discarded.

6.4 A MATTER OF GRAVITY

Professor: Can someone give me the answer?

First Student: According to my pocket calculator, when I substitute the given numbers into the given formula, I obtain a force of 3.335 x 10^{-6}

newtons, or 0.33 dynes. I am suspicious since the problem is so easy, but I am using one of the most basic formulas of physics in a straightforward way, and I am confident that my answer is correct.

Professor: Let's look at that equation again, as applied to this problem, but this time let X be 0.001 meters instead of 0.1 meters. Now the force comes out to be 100 squared or 10,000 times as large as before. What about that? Do you believe it? In fact, the formula says that as X approaches zero, the force approaches infinity. But we know perfectly well, from symmetry considerations, that if the small object is at the center, the net force on it must be zero, not infinity. How do we resolve this discrepancy?

READER: Here is your chance. Then turn to page 94.

READER: Turn page for solution.

Second Student: When Newton showed that the distance from the center of the earth to an object *outside the earth* is the relevant distance for his law to apply, it does not follow that the same principle applies inside a hollow sphere, or shell. As a matter of fact, I can easily prove that the gravitational field exerted by a shell is exactly zero *everywhere* inside the shell.

Our shell is thick-walled. However, consider a spherical shell with a very thin wall, as shown in Figure 6.4B. Imagine a small test mass at some random position within the shell. Now visualize a cone of space extending in any direction from the test mass to the shell, with a small cone angle, θ, as shown.

If X is the distance within the cone from the test mass to the shell, the area of the shell intercepted by the cone is $(X\theta)^2$, if θ is small and is expressed in radians. The mass of the portion of the shell intercepted by the cone is $\delta(X\theta)^2\rho$, where δ is the thickness and ρ the density of the shell.

By Newton's law, the gravitational attraction between the test mass and the portion of the shell intersected by the cone is seen to be independent of

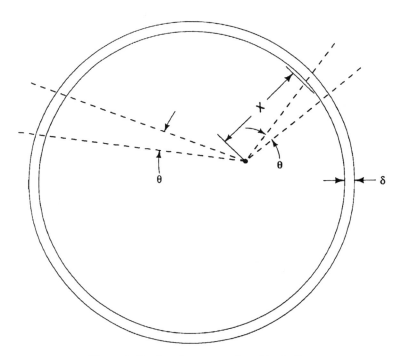

Figure 6.4B   Imaginary cones in a thin shell.

X, since X^2 is in the numerator as well as in the denominator, and therefore is canceled out.

It follows that no matter in what direction the cone is projected from the test mass, the force is the same. If we imagine the cone projecting in exactly the opposite direction from that shown, the force must be equal and opposite, so it cancels out. Thus, there can be no net gravitational force on a test mass inside a thin, hollow sphere.

We have a thick sphere, not a thin sphere. However, a thick-walled hollow sphere may be thought of as a series of thin spheres, one inside the other. Since each of the thin spheres will exert no net force on the test mass inside, this statement must also be true for the assembly of thin spheres, i.e., the thick-walled sphere.

Professor: Excellent. You may be interested to know that there is an analogy to this effect that is encountered when one studies electrostatic fields. The Faraday cage is based on this effect. Clearly, electrostatic fields vary with the inverse square of distance, just as gravitation does, so the same derivation would apply.

6.5 LETTING AIR INTO AN EMPTY TANK

Professor: Who can see a reason why the temperature should be other than 20 degrees C?

First Student: I am familiar with the Joule-Thomson effect, which can cause cooling when a gas expands through an orifice from a high pressure to a low pressure. For example, if the valve on a cylinder containing air at 100 atmospheres and 20 degrees C is opened, the air will emerge well below 0 degrees C, and frost may form on the valve. Perhaps this effect is important in our problem.

Professor: No, the Joule-Thomson effect is not important in our problem because the Joule-Thomson coefficient for air at 1 atmosphere is only 0.23 degrees per atmosphere. That effect is important only at high pressures, when the molecules are originally close together and have to overcome intermolecular attraction when expanding. This requirement causes an expenditure of energy, and the temperature drops. In our problem, this drop would only be a small fraction of a degree. There is a much more important effect in our problem. Who knows what it is?

Second Student: Let's imagine that one-fourth of the air enters the tank first and comes to a pressure of one-fourth of an atmosphere. Then the other three-fourths of the air enters and compresses the original air from 1/4 atmosphere to 1 atmosphere, a fourfold compression. Obviously, such a compression will cause a substantial temperature rise in the gas, unless it has time to lose heat to the tank walls.

But the tank is large, and the whole process is over in 30 seconds, so there is little opportunity for heat transfer to the tank walls during the process, and we can assume adiabatic behavior.

Therefore, I say that the temperature will rise substantially, but I don't know how to calculate the rise since each element of gas is being compressed a different amount, and the gas that is doing the compressing at a given time is itself being compressed at a later time.

Professor: Qualitatively, you are correct. The temperature will rise from 20 degrees C to 137 degrees C, assuming perfect mixing in the tank and no heat loss, and will then gradually cool to ambient temperature, over a period of several hours, as heat transfer takes place.

We can rather easily derive the formula for calculating this temperature rise. We simply write an energy balance. The n moles of air that have entered the tank, before heat loss has occurred, have increased in energy by the quantity $nC_v(T_2-T_1)$. Here, C_v is the heat capacity at constant volume per mole of air, T_1 is the ambient temperature, and T_2 is the maximum temperature of the mixed air in the tank.

Where does this energy come from? The earth's atmosphere, at pressure P_1, has pushed a volume of air, V_1, into the tank. The energy expended by the atmosphere is P_1V_1, which is the mechanical work done when the volume of air, V_1, is pushed into the tank by pressure P_1. Equating this energy to the temperature rise, we have

$$P_1V_1 = nC_v(T_2 - T_1) \qquad\qquad 6.5$$

But by the perfect gas law, with R being the constant,

$$P_1V_1 = nRT_1 \qquad\qquad 6.6$$

Substituting Equation 6.6 into Equation 6.5, we obtain

$$RT_1 = C_v(T_2 - T_1) \qquad\qquad 6.7$$

Rearranging Equation 6.7, we find

$$T_2/T_1 = (C_v + R)/C_v$$

6.8

Those who have studied thermodynamics know that $C_v + R = C_p$, for perfect gases, and $C_p/C_v = \gamma$, the ratio of heat capacity at constant pressure to the heat capacity at constant volume. Thus,

$$T_2/T_1 = \gamma$$

6.9

For air, γ is close to 1.4. Now we can calculate T_2 when $T_1 = 273 + 20 = 293$ K. We find that $T_2 = 410$ K, or 410 - 273 = 137 degrees C.

Notice that Equation 6.9 is yet another example of a situation in which absolute temperature rather than Celsius temperature is in control.

Third Student: You recall the assumption that the gas in the tank is perfectly mixed. It is conceivable to me that the mixing will often be less than perfect, at the end of the 30 seconds. In such a case, couldn't hot spots exist within the tank, so that the *average* final temperature is 137 degrees C, thereby satisfying the conservation of energy, but some local temperatures might be much higher? This fact would be important if we were dealing with a combustible gas, which might ignite, rather than just air.

Professor: That is a very good point. Suppose that after 1/10 of an atmosphere was reached, a pocket of gas that was present was compressed from 1/10 to 1 atmosphere without mixing or heat loss. The equation for temperature rise caused by adiabatic compression is

$$T_2/T_1 = (P_2/P_1)^{(\gamma - 1)/\gamma}$$

6.10

If we insert a value of 10 for the pressure ratio and 1.4 for γ, we find that the final temperature is 566 K, or 293 degrees C. This would be a high enough temperature to cause ignition of some oil vapors in air.

6.6 THE TWO CAPACITORS

Professor: Will someone explain this problem, without violating the principle of conservation of energy?

First Student: In spite of the problem statement, how do I know for sure that the voltage really drops in half when the two capacitors are connected together? If the voltage had dropped to only 0.707 times the original value, it is easy to show that the energy would not change.

Professor: The charge that was originally in the first capacitor was simply a measure of the number of electrons that were in excess on one plate and were deficient on the other plate. When the two capacitors were connected together, obviously the electron excesses and deficiencies would distribute themselves equally. Also the voltages on each side of the pair of capacitors would have to be the same because of the connections. Furthermore, the relationship among the charge, Q, the voltage difference, V, and the capacitance, C, for each capacitor is $Q = VC$; so clearly if the charges after the connections are each half as great as the original charge, the voltage difference across each capacitor must be half as great as before. I am afraid that you are stuck with the original statement of the problem.

Second Student: Even though the connecting cables were said to have low resistance, they still must have finite resistance. Let this resistance be R. When the connection is first made, voltage V_0 is acting across R, and by Ohm's law a current, V_0/R, must flow. Since $V_0 = Q_0/C$, we can also express this initial surge of current as Q_0/RC. It dies down after a short time, but while it is flowing, energy is being dissipated in the resistance at the rate I^2R. This result would explain some energy loss.

However, it would seem that this energy loss would depend on the magnitude of the resistance R, and therefore would not in general be exactly equal to one- half the original stored energy. For instance, if we substitute Q_0/RC for the initial current in the resistive dissipation power term I^2R, we obtain Q_0^2/RC^2 for the initial rate of power dissipation. Obviously it depends on R.

Professor: You are on the right track, but you haven't gone far enough. Consider Figure 6.6A, which represents the state of affairs before and just after closing a switch. At any instant, the charge on the first capacitor is Q_1, which is a decreasing function of time. Meanwhile, the charge on the second capacitor at that instant is Q_0-Q_1, since charge is conserved. The

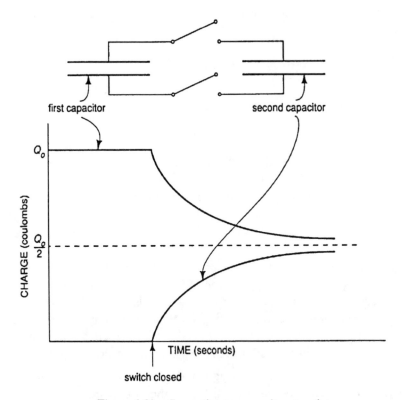

first capacitor

second capacitor

Figure 6.6A   Connecting two capacitors together.

current at that instant is $-dQ_1/dt$. Ultimately, the current decreases to zero and Q_1 decreases to $Q_0/2$.

We can now write and solve a differential equation for this circuit that will give us this expression for the transient current as a function of time t:

$$I = (Q_0/RC)e^{-2t/RC} \qquad\qquad 6.11$$

But the energy dissipated in the resistor during this transient, appearing as heat, is given by

$$\text{Dissipated energy} = \int_0^\infty I^2Rdt \qquad\qquad 6.12$$

If we substitute Equation 6.11 into Equation 6.12, we obtain

$$\text{Dissipated energy} = (Q_o{}^2/RC^2)\int_0^\infty e^{-4t/RC}\, dt \qquad\qquad 6.13$$

When we integrate Equation 6.13, we find, to our surprise and gratification, that the resistance, R, cancels out, and the result is

$$\text{Dissipated energy} = Q_0{}^2/4C \qquad\qquad 6.14$$

Since $Q_0{}^2 = C^2V_0{}^2$, we may rewrite Equation 6.14 as

$$\text{Dissipated energy} = CV_0{}^2/4 \qquad\qquad 6.15$$

Equation 6.15 is valid regardless of how small or how large the resistance is. This dissipated energy is seen to be exactly one-half the original stored energy in the first capacitor, so the paradox is resolved. The missing energy is dissipated as ohmic heating of the connecting cables. If the resistance of these cables were to be reduced to near zero, the initial current flow would be near infinity, so the resistive heating would still be there.

Second Student: A few minutes ago, I showed that the *initial* rate of resistive heating would be smaller if the resistance were larger. How do we reconcile this fact with Equation 6.15, which does not contain resistance?

Professor: If you look at Equation 6.11, you see that time occurs only in the grouping $2t/RC$. Accordingly, the duration of the current transient would be of the order of RC because when t is equal to RC, the exponential term becomes equal to e^{-2}, or 0.135. At that time the current has decayed to 13.5 percent of its peak value. Accordingly, the larger R is the longer the duration of the current surge. This relationship compensates exactly for the fact that the initial rate of dissipation is inversely proportional to R.

CHAPTER 7

VIOLATING THE LAWS OF THERMODYNAMICS

7.1 INTRODUCTION

The first law of thermodynamics insists that energy is always conserved, even though it may appear in many forms: mechanical work, kinetic energy, electrical energy, potential energy (as in a gravitational or magnetic field), chemical energy (e.g., the heat of combustion of methane and oxygen), and even the energy associated with mass-energy transformations (as in nuclear reactions). However, the first law makes it clear that it is not possible to invent a device that generates heat or useful work without depleting energy from some source.

This law was enunciated 140 years ago, and no exceptions have ever been found. Nevertheless, inventors have persisted in proposing devices that allegedly "get around" this law. We shall analyze some of these devices, which are called "perpetual motion machines of the first kind."

The second law of thermodynamics is more subtle and can be stated in various ways. One statement is that it is impossible to construct an engine that, operating in a cycle, will produce no other effect than the extraction of heat from a reservoir and the performance of an equivalent amount of mechanical work. In other words, heat cannot be converted completely into work in a cyclic process. (Notice that this statement does *not* violate the first law.) A device that allegedly converts heat completely into work in a cyclic process is called a "perpetual motion machine of the second kind." Again, no such device is known to exist.

An alternate statement of the second law is that it is impossible to construct a device that, operating in a cycle, will produce no effect other than the transfer of heat from a cooler to a hotter body. It can be demonstrated that this statement of the second law is equivalent to the previous statement.

We shall also analyze some devices that appear to violate the second law but not the first.

7.2 UNDERWATER ROTATING DEVICE

Consider the device shown in Figure 7.2A. It consists of a chain belt looped over two sprocket wheels. Attached to the chain belt is a series of evenly spaced cylinders, each containing a massive, free piston. Each piston is free to move back and forth within its cylinder but does not permit water or air to leak past it. The air has been exhausted from the internal free space in each cylinder, so a vacuum is present in each cylinder.

The two sprocket wheels, one above the other, are supported on shafts that can transmit power to an external system if rotation occurs. The entire device is submerged in water.

The weight of each free piston is sufficient so that when the open end of the cylinder is turned downward, as on the right hand side of the drawing, the piston will move to the bottom of the cylinder, as shown. The weight of the piston overcomes the combined atmospheric and hydrostatic pressures and forces water out of the cylinder. When the open end is turned upward, the reverse occurs.

Let's focus attention on any pair of cylinders on the same level. The one on the right will contain a free volume, V, whereas no such volume in present in the one on the left. Accordingly, although the masses of the two cylinders are the same, there is a net buoyancy force equal to $gV\rho_w$ on this pair of cylinders, which will tend to cause the chain belt to rotate counterclockwise. (Here, ρ_w is the density of water, and g is the local acceleration of gravity.)

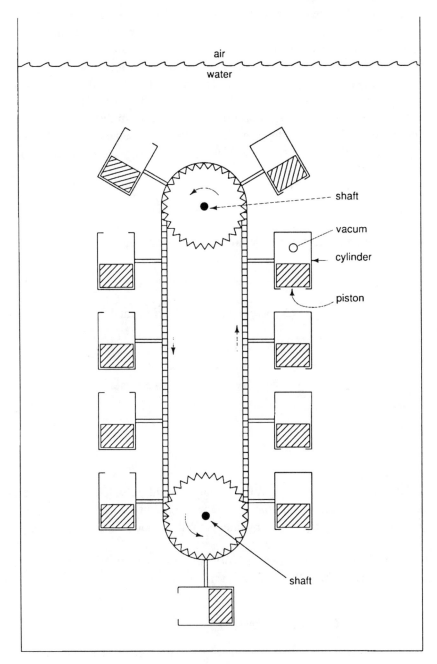

Figure 7.2A Underwater rotating device.

A similar force will act on each pair of cylinders. Therefore, one would expect rotation in a counterclockwise direction to result. As each cylinder passes over the top of the upper sprocket wheel, its piston should shift from one end of the cylinder to the other. A similar action should occur at the bottom. Meanwhile, power would be expected to be transmitted to the sprocket wheel shafts and could be used externally, if friction is small.

We can calculate how much this power might amount to. If n is the number of cylinders, t is the time for a cylinder to make a complete round trip, and h is the height from the bottom to the top, we see that the power (force × distance/time) is:

$$\text{Power} = n(gV\rho_w)(h)/t \qquad\qquad 7.1$$

Let's insert some reasonable numbers into this equation:

$$n = 20$$

$$g = 9.8 \text{ meters per second per second}$$

$$V = 0.1 \text{ cubic meters}$$

$$\rho_w = 1{,}000 \text{ kilograms per cubic meter}$$

$$h = 5 \text{ meters}$$

$$t = 10 \text{ seconds}$$

We then find that the power generated would be 9,800 watts. Accordingly, we have a machine that will generate about 10 kilowatts of power, forever, without consuming any energy. Will it work? If not, why not?

READER: If you believe it will work, build one yourself. If you do not believe, what is the fallacy? Then turn to page 111.

7.3 ELECTROLYSIS OF WATER

Imagine a steep cliff, with an electrolysis unit at the base of the cliff, capable of dissociating water into hydrogen and oxygen, and a fuel cell at the top of the cliff, capable of combining hydrogen and oxygen into water while generating electricity. The situation is shown in Figure 7.3A.

The water generated at the top flows down a pipe to the electrolysis unit, which is 100 meters below. At the discharge end of the pipe, the water pressure will be about 10 atmospheres because of the hydrostatic pressure of a column of water 100 meters high. Accordingly, this water can be put through a water turbine, which generates useful power, before the water enters the electrolysis unit at atmospheric pressure.

The hydrogen and oxygen generated at the bottom can be gotten back up to the top without the expenditure of energy because the hydrogen is much lighter than air and could be put into balloons, which float up to the top. The oxygen is only slightly heavier than air and could also be put into balloons, which are pulled to the top by the hydrogen balloons, especially since there are two volumes of hydrogen for each volume of oxygen.

Although neither the fuel cell nor the electrolysis unit can operate at exactly 100 percent efficiency, there is no thermodynamic reason why their efficiencies could not each be close to 100 percent. If we assume that their efficiencies were each very close to 100 percent, it would appear that the electricity generated at the top would be exactly what is required to power the electrolysis operation at the bottom since the chemical reactions at the top and the bottom are exactly the reverse of one another.

The net result, then, would be the power generated by the water turbine. Some of this power might be used to make up for the small inefficiencies of the fuel cell and the electrolysis unit, and the remainder would be available for any desired use. No raw materials would be consumed.

READER: Could this scheme possibly work? If not, why not? Then turn to page 113 for the solution.

7.4 A SUBMARINE THAT NEEDS NO FUEL

Heat-transfer coils are mounted on the surface of a submarine. A cold, refrigerant-type fluid is circulated through these coils and picks up heat from the ocean. Its temperature rises to a value close to that of the ocean. The heat acquired by the fluid is transferred to an engine within

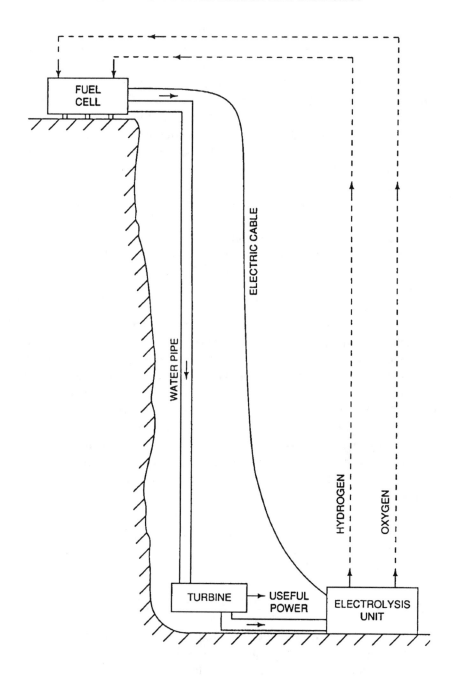

Figure 7.3A   Power generation scheme involving electrolysis of water.

the submarine that converts this heat into work by heating a gas that expands and pushes a piston. During expansion, the gas cools to its original temperature and the cycle can be repeated. Meanwhile, the refrigerant-type fluid, having given up heat to the gas, has cooled and must be again heated to near the ocean temperature by being recirculated through the coils on the surface of the submarine. The cycle repeats indefinitely. The submarine is propelled by the piston movement, transmitted to a propeller. No fuel is consumed.

We have not violated the principle of conservation of energy because the energy to run the submarine has been taken from the heat in the ocean water.

READER: Why won't this system work? Then turn to page 115 for the solution.

7.5 STEAM AND ETHYLENE GLYCOL

A friend told me that she had demonstrated that heat could be transferred from a lower to a higher temperature. She said that she had performed the following experiment: She started with an open vessel, insulated on the sides and bottom, containing 1 kilogram of ethylene glycol (the familiar antifreeze liquid) at 20 degrees C. She inserted a tube into this vessel, through which she injected a constant flow of steam at 100 degrees C. She observed that some of the steam condensed and dissolved in the ethylene glycol since the volume of the liquid gradually increased. A thermometer was immersed in the liquid. (The setup is shown in Figure 7.5A.) As the process continued, the temperature rose from 20 degrees C and ultimately reached 130 degrees C.

My friend's first thought was that some sort of exothermic reaction must be occurring between the steam and the ethylene glycol. However, upon studying the literature, she found that the ethylene glycol-water system comes fairly close to forming an "ideal solution." That is, negligible heat is evolved or absorbed when water and glycol are mixed, they are mutually soluble in all proportions, and the volumes are additive.

In view of the failure of this explanation, my friend asked me how she could reconcile her result with the principle that heat cannot flow from steam at 100 degrees C to a liquid at a higher temperature.

She provided some facts that she thought might be helpful.

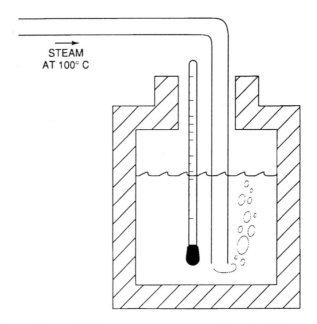

Figure 7.5A   Bubbling steam at 100 degrees C through ethylene glycol.

1. The pressure was exactly 1 atmosphere.

2. Water boils at 100 degrees C and ethylene glycol boils at 197 degrees C.

3. The heat capacity of liquid water is 4.18 joules per gram per degree, and that of ethylene glycol is 2.4 joules per gram per degree.

4. The molecular weights of water and ethylene glycol are 18 and 86, respectively.

5. The heat of condensation of steam is 2,255 joules per gram at 100 degrees C.

6. The vapor pressure of water is 1.00, 1.41, 1.96, and 2.67 atmospheres at 100, 110, 120, and 130 degrees C, respectively.

READER: Do you believe that the temperature rose to 130 degrees C? If so, can you explain it on the basis of the preceding facts? And can you reconcile it with thermodynamics? Turn to page 119 for the solution.

7.6 AN ODD CASE OF RADIATIVE EXCHANGE

An ellipse has two foci. If we draw a straight line from one focus to any part of the ellipse, so that the line intercepts the ellipse at some angle to the normal, and then draw a second line from the interception point that is at the same angle on the other side of the normal, the second line will pass through the other focus. This is a geometric fact.

If we rotate an ellipse about its major axis, it forms a prolate spheroid, the two foci of which are the same as the foci of the ellipse. Again, a ray from either focus that reflects specularly from anywhere on the inside surface of the spheroid will pass precisely through the other focus.

Now, let's suppose that we construct a hollow object such as shown in Figure 7.6A. Imagine two prolate spheroids, one inside the other, but both with the same pair of foci, as shown. One end of the hollow object we have built coincides with the smaller spheroid, and the other end coincides with the larger spheroid. The transition section is part of an imaginary sphere the center of which is one of the foci, as shown. The inside of this hollow object is highly polished and perfectly reflective.

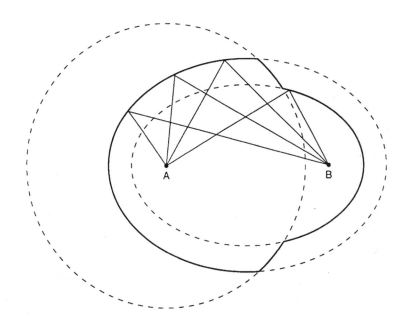

Figure 7.6A Radiative exchange between two spheres, A and B, inside a perfectly reflective enclosure consisting of parts of two spheroids.

Finally, we place hot, black spheres A and B at the two foci, as shown, and evacuate the air. All the radiation from sphere B is seen to be reflected to sphere A, and most but not all the radiation from sphere A is reflected to sphere B. Part of the radiation from sphere A will impinge on the spherical transition section and will come right back to sphere A.

Even though A and B were originally at the same temperature, A must get hotter and B must get cooler since all the radiative flux from B goes to A and only some of the flux from A goes to B.

This phenomenon will permit us to operate a heat engine cycle by absorbing heat from A into the engine, letting the engine work, and transferring the residual heat (at a lower temperature) along with some externally supplied heat (equivalent to the work done) to B. This process can be continued indefinitely. We are converting heat into work without rejecting any heat, which is contrary to the second law of thermodynamics. We are also transferring heat from a lower to a higher temperature, by radiation, which should not be possible.

READER: We all know that this process won't work. But why not? Then turn to page 121.

SOLUTIONS TO CHAPTER 7 VIOLATING THE LAWS OF THERMODYNAMICS

7.2 UNDERWATER ROTATING DEVICE

Professor: I take for granted that none of you thinks that this device will work. But why not?

First Student: I can write the first law of thermodynamics as

$$\Delta E = Q - W \qquad\qquad 7.2$$

where ΔE is the internal energy change, Q is the heat added to the system, and W is the work done by the system. In this case, both ΔE and Q are zero. Hence, W must be zero. I rest my case.

Professor: Yes, but what is wrong with the argument presented? I assure you that the buoyancy force was calculated correctly. Is there an opposing force, and if so, what?

Second Student: I would say that friction is the culprit. As these cylinders move through the water, their motion must be resisted by friction, which evidently is sufficient to prevent the rotation.

Professor: I don't accept that explanation. In the first place, friction would be proportional to the cross-sectional area of the cylinders, and the buoyancy is proportional to the volume of the cylinders. The volume-to-surface ratio is proportional to the size. If we make the cylinders large enough, the buoyancy force will overwhelm the friction force.

In the second place, the frictional force is zero at zero velocity and increases with increasing velocity. There is obviously a finite velocity at which the friction force becomes equal to the buoyancy force (analogous to the terminal velocity of a released helium-filled balloon). Accordingly, there would always be a finite rate of production of power at a sufficiently low velocity, unless something else is involved.

First Student: What about the shifting of position of those massive pistons at the top and bottom? Inertial forces must be involved as those pistons slide in their cylinders. We can't assume that the pistons are very light because they must be heavy enough to overcome the external pressure. I suspect that this is the culprit.

Professor: I concede that there are inertial effects associated with these masses. But consider the following argument: We could imagine that the device is 100 meters high instead of 5 meters high, as was originally assumed, and we have 100/5 = 20 times as many cylinders attached to the chain belt. If the belt moves at the same speed, the buoyancy force is 20 times as large, whereas the inertial effect of shifting weights at the top and the bottom is the same. Accordingly, such an effect cannot stop the rotation. Furthermore, the piston at the top shifts in a counterclockwise direction, whereas the shift at the bottom is clockwise, so the inertial effects of these shifts should cancel one another.

Third Student: Let P_t be the pressure (atmospheric plus hydrostatic) at the top of the device and P_b the pressure at the bottom, and h is the distance between top and bottom. Then, when a cylinder has just passed the lowest point and is starting to rise on the right-hand side, its piston must work to expel the water from the cylinder and create the empty volume, V. This work is equal to $P_b V$, and is work done by the device on the water. At the top, the water does work on the device when each piston moves the other way; this work is equal to $P_t V$. Since P_b is larger than P_t, the net work equal to $(P_b - P_t)V$ is done by the device on the water each time a cylinder goes around. Let n be the number of cylinders and t the time for a cylinder to go around. Then, the power expended for this pumping of water by the pistons is

$$\text{Power} = n(P_b - P_t)V/t$$

7.3

But according to hydrostatics,

$$P_b - P_t = gh\rho_w$$

7.4

where ρ_w is the density of water. Substituting Equation 7.4 into Equation 7.3, we obtain

$$\text{Power} = ngh\rho_w V/t \qquad\qquad 7.5$$

which is exactly the same as Equation 7.1, for the power generated by buoyancy. Accordingly, the device can produce *no net power*, quite aside from inertial and friction forces.

Professor: Right on the target. Notice that the explanation of the fallacy becomes overwhelmingly convincing when an exact calculation can be made that shows how the buoyancy is exactly counterbalanced by the pumping effect.

7.3 ELECTROLYSIS OF WATER

Professor: Who sees any flaws in this scheme if we concede that high-efficiency fuel cells and electrolysis units can be developed?

First Student: It seems to me that the presentation was too conservative. There is no need to carry the oxygen generated by the electrolysis unit up to the fuel cell. There is plenty of oxygen in the air at the top of the cliff. So all the oxygen that is generated can be sold.

Further, as the hydrogen rises to the top in the balloons, there is a net buoyancy force here that is not being used. These hydrogen balloons could be attached to some sort of device such as we had in the previous problem, and useful energy could be obtained from this device in addition to the energy obtained from the water turbine.

Of course, thermodynamic principles rule out the possibility that this scheme could work, with or without my additions, but I don't see why.

Second Student: The electricity generated at the top must be conducted to the bottom, and this cannot be done without transmission losses, in the form of ohmic resistance. There is also a frictional loss as the water flows down the pipe. When these losses are considered, perhaps it will be clear why the scheme cannot work.

Professor: In regard to ohmic resistance and pipe friction, there is no reason why we cannot use large-diameter copper conductors and a large-diameter pipe, and therefore reduce these losses to as small a value as we wish.

Third Student: I have studied electrochemistry, and I see the basic fallacy in the scheme. The electrical energy ideally generated by the fuel cell is equal to the reduction of the free energy when hydrogen and oxygen gases change into liquid water. Similarly, the electrical energy ideally required to dissociate liquid water into gaseous hydrogen and oxygen is also equal to the free energy difference between reactants and products. Free energies depend on temperature and pressure. We can assume that the temperature is the same at the top and the bottom of the cliff, but most certainly, the pressure is not the same. When the pressure of a gas is changed from P_1 to P_2, its free energy change, ΔF, is given by

$$\Delta F = RT ln(P_2/P_1)$$

where R is gas constant and T is absolute temperature.

Accordingly, since the pressure at the top is lower, the free energies of the hydrogen and oxygen at the top are lower, and the voltage generated by the fuel cell at the top must be lower than the voltage required to dissociate water at the bottom. Under ideal, frictionless conditions, the extra energy that may be generated by the downward and upward transport of liquid water and gases can be shown to be just equivalent to this voltage deficiency. It is obvious that making the cliff twice as high would double the energy available from the transport process, but it would also double the value of $ln(P_2/P_1)$, according to the equation for the variation of pressure with altitude, for an isothermal atmosphere:

$$1n(P_2/P_1) = g(h_1 - h_2)(M/RT)$$

where g is the acceleration of gravity and M is the molecular weight of air. I conclude that no useful power can ever be produced by such a scheme.

First Student: But what about my suggestion that we could manufacture oxygen for sale in this way, without expending any energy, under frictionless, ideal conditions?

Third Student: Since the partial pressure of oxygen in air is only 21 percent as great as in pure oxygen, the free energy in the oxygen of the air is lower than for pure oxygen; thus the voltage generated by a hydrogen-air fuel cell is lower than that of a hydrogen-oxygen electrolysis unit even with both at the same altitude. In summary, no free lunch.

7.4 A SUBMARINE THAT NEEDS NO FUEL

Professor: We are all in agreement, I am sure, that this propulsion method will not work, but who can explain just why not?

First Student: Although we are not violating the first law of thermodynamics, we are clearly violating the second law. According to one well-known formulation of the second law, the ratio of the work, W, produced by a heat engine to the heat input, Q_1, is always less than a certain value depending on absolute temperatures:

$$W/Q_1 < [1 - (T_2/T_1)] \qquad\qquad 7.6$$

Here, T_1 is the absolute temperature at which heat is added, and T_2 is the absolute temperature at which the unused portion of the heat is withdrawn. Since T_2 is never absolute zero and T_1 is never infinity, the equation insists that W/Q_1 is always less than 1. We know from the first law that $W = Q_1 - Q_2$, so clearly Q_2 is always finite. In fact, if we combine the first law with Equation 7.6, we obtain

$$Q_2 > Q_1 T_2/T_1 \qquad\qquad 7.7$$

In this problem, the submarine is immersed in the ocean, which is at temperature T_1. If the submarine absorbs some heat, Q_1, from the ocean and converts some of it to work, the remainder, Q_2, must be rejected to a heat sink at a lower temperature T_2. But this process is obviously impossible since the submarine is surrounded by water at temperature T_1.

However, I must say that even though I know how to write Equations 7.6 and 7.2, I don't feel that I understand them.

Professor: Your exposition was perfectly correct. The second law of thermodynamics is a basic law, like the law of gravitation, the laws of motion, and Maxwell's equations for electromagnetic waves. We accept these laws if experimentalists cannot find exceptions to them. So don't say you don't "understand" the second law of thermodynamics.

However, can anyone find other words to explain why this submarine-propulsion scheme won't work?

Second Student: We can analyze the problem in terms of entropy. If the submarine absorbs heat Q_1 from the ocean at temperature T_1, there is an entropy flow of magnitude Q_1/T_1 into the submarine. *If* the submarine had access to a lower-temperature heat sink at temperature T_2, it could reject heat Q_2 into this heat sink; this action would correspond to rejecting entropy of magnitude Q_2/T_2. According to Equation 7.7, slightly rearranged,

$$Q_2/T_2 > Q_1/T_1 \qquad\qquad 7.7$$

This formula says that the second law requires the entropy rejected to be greater than the entropy taken in originally. There is a net increase in entropy. (If the engine were ideal, i.e., frictionless and perfectly insulated, the entropy rejected could be the same as the entropy taken in, and there would be no net change in entropy.)

However, in no circumstances could there be a net *decrease* in entropy. We can in some sense understand this statement by equating an entropy increase with a transition from a more ordered to a less ordered or more random state, which is a more probable state.

In the case of the submarine, there is no entropy, or randomness, associated with the propulsion work, W. If the heat input, Q_1, could be entirely converted to work, the input entropy, Q_1/T_1, would have disappeared, and there would be a net reduction of entropy, which is not possible. On the other hand, since the surroundings are at T_1, it is not possible for unconverted heat to be rejected at a temperature lower than T_1. We could get around the difficulty if we could reject heat at a temperature T_2 that is higher than T_1. But we can't make heat flow from a lower to a higher temperature.

Possibly, the fallacy of this propulsion scheme is more understandable when it is discussed in terms of entropy.

First Student: Maybe so. But I would feel more comfortable if we went back to the original description of the heat-power cycle, which sounded plausible, and pinpointed the reason it wouldn't work, without invoking entropy. Could this be done?

Professor: Certainly. Let's go over the proposed cycle, using some assumed numerical values. Let the ocean temperature be 290 K. Assuming perfect heat transfer, the circulating fluid picks up a quantity of heat Q_1 from the ocean at temperature T_1 and transfers it to some helium in a cylinder-piston arrangement. The helium is assumed to be initially at a lower temperature, say 240 K, and is heated at constant volume to 290 K by the circulating fluid. This heating causes its pressure to rise from an assumed initial value of 1 atmosphere to a value 290/240 times as high, which is 1.21 atmospheres.

Then we let the helium expand adiabatically back to 1 atmosphere, pushing the piston frictionlessly and generating mechanical energy. During this adiabatic expansion from P_1 (1.21 atmospheres) to P_2 (1 atmosphere), the temperature decreases from T_1 (= 290 K) to T_2 according to the thermodynamic relationship

$$T_2/T_1 = (P_2/P_1)^{(\gamma - 1)/\gamma}$$

7.8

where γ, the specific heat ratio, is 5/3 for helium. When we calculate the temperature after expansion from this formula, we obtain T_2 = 269 K. Notice that the temperature did not come back to the original assumed value of 240 K.

At this condition, the volume occupied by the helium is 269/240 = 1.12 times its original volume. If we could now cool the helium to a lower temperature, so that its volume would contract to its original volume, we could then proceed to a repetition of the previous steps. However, we do not have a heat sink cold enough to do so.

If we leave the volume as it is and heat the helium back to 290 K, the pressure will increase from 1 atmosphere to only 1.078 atmospheres. If we then expand back to 1 atmosphere, extracting power, the temperature then drops to only 281 K, according to Equation 7.8. The volume is now 1.17 times the original volume.

A continuation of this calculation is shown in Figure 7.4A. We see that in a short time the pressure and temperature changes become insignificantly small, and essentially no more propulsion energy is available.

First Student: That helps a lot. But now I am curious about how you could prove that the cycle doesn't work without using the second law of thermodynamics at all.

Professor: As a matter of fact, Equation 7.8, which is called the law of isentropic compression or expansion of a perfect gas, is derived from the second law of thermodynamics.

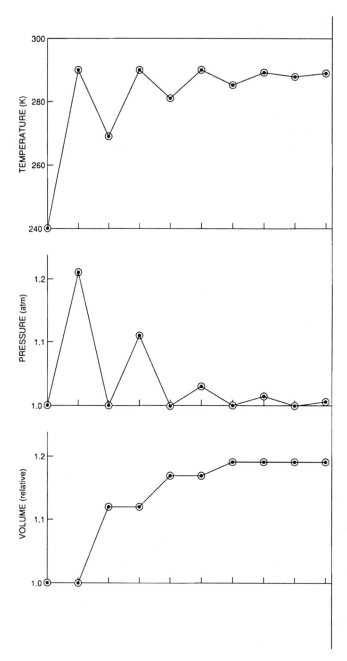

Figure 7.4A   Calculated temperature, pressure, and volume history when helium is alternately heated at constant volume and expanded adiabatically.

7.5 STEAM AND ETHYLENE GLYCOL

Professor: Here we go again, trying to get around thermodynamic principles. Who has an opinion on this one?

First Student: If we accept that no significant exothermic chemical combination of steam and ethylene glycol is occurring, it seems perfectly clear that the condensing steam will heat the liquid to 100 degrees C, after which nothing more will occur. This explanation makes sense and is consistent with thermodynamics. If the experiment gave a higher temperature, perhaps the steam being used was superheated, or the thermometer was wrong.

Professor: I am afraid that you are not aware of Raoult's law for ideal solutions. This law says that the vapor pressure of water over a water-ethylene glycol solution is equal to the product of the mole fraction of water in the solution and the vapor pressure of pure water at the temperature of the solution. As a consequence of this law, the initial boiling point of a water-ethylene glycol solution is higher than 100 degrees C.

At the reported temperature of 130 degrees C, the vapor pressure of water is 2.67 atmospheres, we were told. It follows that a glycol-water solution containing a mole fraction of water equal to 1/2.67 = 0.37 would have a vapor pressure of 1 atmosphere. At any temperature below 130 degrees, the vapor pressure over the solution would be less than 1 atmosphere, and steam inside a bubble at 1 atmosphere would condense and dissolve in the solution.

It remains to be seen if the facts are consistent with the mole fraction of water in the solution being 0.37 when the temperature reaches 130 degrees C. Who knows how to calculate this figure?

Second Student: We started with 1,000/86 = 11.62 moles of ethylene glycol. The molar ratio of water to glycol in the final solution would be 0.37/(1 − 0.37) = 0.59. Accordingly, we have 11.62 × 0.59 = 6.86 moles of water, or 6.86 × 18 = 123 grams of water.

This result means that 123 grams of steam have condensed, liberating 2,255 joules per gram. Therefore, 277,000 joules have been liberated.

The heat required to bring 1,000 grams of glycol from 20 to 130 degrees is 2.4 × 1000 × (130-20) = 264,000 joules. Some heat is needed to heat the condensed steam from 100 to 130 degrees; this is 4.18 × 123 × (130 − 100) = 15,400 joules.

Thus, the total heat requirement is 264,000 + 15,400 = 279,400 joules, and the heat available from the condensation energy is 277,000 joules. I'd say that this is just about perfect agreement.

First Student: I follow the calculations and see that the first law of thermodynamics is obeyed, but I still feel that the second law is being violated.

Professor: Consider a hot reservoir at T_1 and a cooler reservoir at T_2. Imagine an engine that extracts heat Q_1 from the hot reservoir, generates mechanical work W, and then rejects heat $Q_1 - W$ to the cold reservoir. This process is consistent, so far, with both the first and second laws. But now suppose that there is some way that the quantity of heat $Q_1 - W$ could be transferred from the cold reservoir back to the hot reservoir. Then we could add an additional quantity of heat equal to W to the hot reservoir, and the net effect would be that we are converting heat into work without rejecting heat. We could continue this process indefinitely by repeating the cycle. However, because this operation is forbidden by the second law, it follows that transferring heat from the cold reservoir to the hot reservoir in such a cyclic operation cannot be possible.

In our glycol-water problem, we have not violated the second law because the water dissolved irreversibly into the glycol, at one time only. There is no way we could make this into a cycle without introducing additional energy to separate the water and the glycol.

Rudolf Clausius, in the nineteenth century, stated the second law in the following language: "It is impossible to construct a device that, *operating in a cycle*, will produce no other effect than the transfer of heat from a cooler to a hotter body."

So much for thermodynamic generalities. We still need to understand why steam at 100 degrees would want to condense onto the surface of a water-glycol solution at, say, 120 degrees. Consider the surface, at a molecular level. Water molecules are evaporating at some rate into the vapor phase, while water vapor molecules are colliding with the surface and condensing at some rate. (The glycol molecules are only evaporating very slightly at 120 degrees, because the boiling point of pure glycol is 197 degrees, so we ignore their evaporation.)

If this were a pure liquid water surface at 120 degrees and 1 atmosphere, the rate of evaporation would be appreciably greater than the rate of condensation, and the net effect would be evaporation. However, in the case under consideration, only about one-third of the molecules on the surface of the liquid are water molecules, the rest being glycol molecules. Therefore, the rate of evaporation of water is only about one-third that of pure water at 120 degrees. On the other hand, the rate of condensation of steam is exactly the same. The net effect is that the rate of condensation is greater than that of evaporation, and this state of affairs continues until the temperature rises to 130 degrees.

It is interesting to note that even though there may a small chemical attraction between water molecules and glycol molecules in excess of the attraction between two water molecules, such an effect would be in addition to what we are talking about and is not required to explain the phenomenon. Indeed, we are able to calculate the 130-degree maximum without knowing anything about a chemical attraction between water and ethylene glycol.

7.6 AN ODD CASE OF RADIATIVE EXCHANGE

Professor: Who is going to tell us the fallacy in this scheme?

First Student: In the presentation, we were told that the same pair of foci served for both the small ellipse and the large ellipse. I am not so sure that this is possible.

Professor: It is perfectly possible. If you draw a straight line from one focus of an ellipse in any direction, and extend it until it intersects the ellipse, and then draw a second straight line to the other focus, you will find that the sum of the lengths of these two lines is the same, no matter what the original direction is. In fact, stick two pins in a sheet of paper to represent two foci; drape a loose loop of string over the pins; use a pencil to pull the string taut; and move the pencil around the pins, maintaining tautness. This process produces an ellipse. If we maintain the pins but repeat with a larger loop of string, we get a larger ellipse with the same foci. That should take care of your concern.

Second Student: We assumed that the inside of the cavity was perfectly reflective, and all the reflections were specular, not diffuse. This assumption would have to include infrared as well as visible radiation. Could such a perfect mirrored surface be produced?

Professor: Obviously, we can never achieve perfection. But I know of no physical principle that would prevent us from coming very close to perfection, if we took enough trouble. So I don't think you can use that argument as a proof that the device would not work.

Second Student: I'll try again. The diagram does not show how the heat is collected from sphere A for use by the heat engine and how the heat at lower temperature is returned to sphere B. Anything that we would insert for this purpose would interfere with the optics.

Professor: Consider this argument: I let the apparatus, as shown, operate for a while, so that sphere A can get hotter than sphere B. Then I remove the two spheres and place them on each side of an external heat engine. After heat exchange takes place, I return the spheres to the enclosure and let the process repeat.

Third Student: Finally, I see what the fallacy is. The two spheres, A and B, must be of finite size. If they were infinitesimal points and the mirrored surfaces were perfect, it is true that all the radiation from point B would go to point A. But to have any hope of transferring a finite amount of heat, A and B must be of finite size. Any radiation beam issuing from sphere B that is not on a line extendable precisely through the center of B (the focal point) may miss A entirely. This radiation may ultimately come back to sphere B. Thus, the whole argument collapses.

CHAPTER 8

SOME PROPAGATION PROCESSES

8.1 A WEAK SHOCK WAVE

A gas is at rest in a tube having a piston at one end. The piston is suddenly pushed a short distance into the tube. This action causes a weak shock wave to propagate through the gas.

Can you deduce a formula for the speed of the shock wave, in terms of the properties of the gas?

Hint: As a gas is compressed adiabatically and reversibly, its pressure at any instant is proportional to its density raised to the power γ (the specific heat ratio).

READER: Can you solve the problem? Then turn to page 126.

8.2 MELTING THROUGH ICE

I have a cube of ice 0.5 meters on an edge, at zero degrees C. I stretch a steel wire, 1 millimeter in diameter, horizontally across the top of the cube. I pass a sufficiently large electrical current through the wire so that 1,000 watts are dissipated per meter of wire length, by ohmic heating. I apply force to the wire and press it downward onto the ice.

123

Eventually, the wire will melt its way downward through the ice, and cut it into two pieces. Estimate how long this process will take.

Here are some properties that may be helpful: The thermal conductivities of ice and liquid water are 2.22 and 0.56 watts per meter per degree C, respectively. The latent heat of fusion of ice is 333 joules per gram. The density of ice is 0.91 grams per cubic centimeter. The melting point of ice decreases about 0.0008 degrees C for each atmosphere increase in pressure.

READER: Can you make an estimate? Then turn to page 130.

8.3 COMBUSTION WAVE IN A THERMITE MIXTURE

A mixture of aluminum powder (Al) and iron oxide powder (Fe_3O_4) is called thermite. If the mixture is formed into the shape of a long cylinder and one end is exposed to heat, "ignition" may result, and a very hot reaction zone will propagate at a steady rate to the far end of the cylinder. The rate of propagation will be several millimeters per second.

This chemical reaction will have occured:

$$8 \text{ Al} + 3 \text{ Fe}_3\text{O}_4 \rightarrow 4 \text{ Al}_2\text{O}_3 + 9 \text{ Fe}$$

This is a highly exothermic reaction, liberating about 3,700 joules per gram of reactants. The reaction products will be at a temperature of about 2,400 degrees C and will be molten. However, no gaseous products form, so no flame is present.

What controls the rate of spread of such a process? Presumably, thermal conductivity is involved, but is the spread rate directly proportional to conductivity? What other properties might it depend on? How would you calculate the energy requirement to initiate such a process?

READER: What are your thoughts? Then turn to page 132.

8.4 FLAME PROPAGATION IN A COMBUSTIBLE GAS

Suppose we have a stationary mixture of 10 percent methane and 90 percent air in a horizontal glass tube 1 centimeter in diameter and several meters long. We open both ends and ignite one end. We observe that a blue flame propagates through the tube at roughly 1 meter per second.

In another experiment, we allow a mixture of 10 percent methane and 90 percent air to flow steadily through a vertical tube at a velocity of 1 meter per second. The mixture emerges from the top of the tube. We ignite the emerging gas and see that a steady blue flame of roughly conical shape sits on the top of the tube. If we increase the gas velocity feeding this flame, the flame becomes taller. If we continue to increase the gas velocity, eventually the flame lifts off the tube and disappears. On the other hand, if we reduce the gas velocity very much below the initial value, the flame "flashes back" into the tube.

Does the propagation rate of a gas flame follow the same laws as the thermite propagation discussed in the previous section? What would be the essential differences, if any?

READER: Do you have any thoughts on this problem? Then turn to page 140.

SOLUTIONS TO CHAPTER 8
SOME PROPAGATION PROCESSES

8.1 A WEAK SHOCK WAVE

Professor: Does anyone remember how to solve this problem, which you undoubtedly encountered in a physics course?

First Student: I know that a gas consists of molecules moving in random directions at various high velocities. Collisions between molecules occur frequently. The motion of the piston will perturb the molecules adjacent to it, and this effect can be transmitted to more remote molecules as the molecules collide with one another.

Accordingly, the disturbance, which we may call a shock wave, cannot move through the gas any faster than the molecular speed, unless the piston velocity is large compared with the molecular speed. Since at any temperature the molecules are moving with varying speeds, let's talk about the most probable speed that a given molecule may have.

"I know from the kinetic theory of gases that this speed is $\sqrt{[2RT/M]}$, where R is the gas constant, T is absolute temperature, and M is the molecular weight. For air at 20 degrees C, this speed is 410 meters per second. I would predict that the shock wave would move through the gas at about this speed, unless the piston's maximum velocity was so great that the molecules acquired significant additional velocity from the piston.

Professor: What you say is quite correct. But is there another way we can approach this problem, which doesn't require knowledge of the kinetic theory of gases?

Second Student: Yes. We don't even need to know that the gas consists of molecules. We simply need to recognize that the shock wave acts to compress (change the volume of) any unit mass of the gas it reaches, and this action requires the unit mass of gas to move, relative to its initial position in the tube. Thus, acceleration occurs, caused by a pressure differ-ence, in accordance with Newton's second law of motion.

We are analyzing a *weak* shock wave. Assume that a small density increase, small enough to be considered a differential quantity, $d\rho$, moves through the gas from left to right at a constant speed, v, relative to the tube. To simplify the analysis, let's use a coordinate system that moves from left to right with velocity v. Then the shock wave is stationary in the coordinate system.

Relative to this coordinate system, the velocities are as shown in Figure 8.1A. The velocity decreases from v to $v + dv$ across the transition zone (dv being negative), while the density increases from ρ to $\rho + d\rho$ across this zone. Since the mass flow entering the transition zone per unit cross section, equal to ρv, must equal the mass flow leaving the zone, it follows that

$$d(\rho v) = \rho dv + v d\rho = 0 \qquad\qquad 8.1$$

By Newton's second law of motion, the change of velocity, dv, is related to the pressure difference, dp, across the transition zone by the expression

$$p - (p + dp) = (\rho v)dv \qquad\qquad 8.2$$

If we eliminate dv between these two equations, we obtain

$$dp/d\rho = v^2 \text{ or } v = \sqrt{[dp/d\rho]} \qquad\qquad 8.3$$

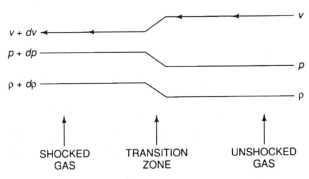

Figure 8.1A   Gradients across a weak shock wave (coordinate system moving with wave).

Now all we need to know is how the density of a gas changes when we change its pressure. For a small amount of rapid compression, we may assume that the process is adiabatic and reversible (isentropic). Then for a perfect gas, the following relation holds:

$$p = C\rho^\gamma \qquad 8.4$$

where C is a constant and γ is the specific heat ratio. Differentiation of Equation 8.4 gives

$$dp/d\rho = C\gamma\rho^{\gamma-1} = C\gamma\rho^\gamma/\rho = \gamma p/\rho \qquad 8.5$$

But by the perfect gas law,

$$p = R\rho T/M \qquad 8.6$$

where R is the gas constant, T is the absolute temperature, and M is the molecular weight of the gas. Substituting Equation 8.6 into Equation 8.5, we obtain

$$dp/d\rho = \gamma RT/M \qquad 8.7$$

Combining Equations 8.3 and 8.7, we get the desired formula:

$$v = \sqrt{\gamma RT/M} \qquad 8.8$$

The speed of propagation, v, is seen to depend on the temperature, the molecular weight, and the specific heat ratio of the gas, but not on the pressure.

Professor: How does this derived speed of a weak shock wave compare with the most probable speed of a gas molecule at the same temperature and same molecular weight?

First Student: We take the ratio of the shock wave speed as shown in Equation 8.8 to the previously mentioned value, $\sqrt{[2RT/M]}$, from the kinetic theory of gases. The ratio is equal to $\sqrt{[\gamma/2]}$. For monatomic gases such as helium or argon, $\gamma = 5/3$ and $\sqrt{[\gamma/2]} = 0.91$. For diatomic gases such as oxygen or nitrogen (comprising air), $\gamma = 1.4$ and $\sqrt{[\gamma/2]} = 0.84$. Thus, the weak shock wave is moving only a little slower than the most probable molecular velocity.

Professor: This discussion has been limited to weak shock waves, and it also applies to ordinary sound waves. The treatment was based on treating the pressure, density, and velocity changes as differentials. It may be of interest to inquire how a strong shock wave would propagate, such as would be produced by a piston moving much faster than the molecular speed.

A different analysis would be required, which would use the preceding equations (conservation of mass, Newton's second law, and the perfect gas law) and also an additional equation for conservation of energy, as well as the assumption that the specific heat of the gas is independent of temperature. The derivation may be found in gas dynamics textbooks. The result is

$$[v_s/v_w]^2 = 1 + [(\gamma + 1)/(2\gamma)][\Delta p/p_o] \qquad\qquad 8.9$$

where v_s is the velocity of a strong shock wave driven by pressure difference Δp, and v_w is the velocity at which a weak shock wave or a sound wave would have propagated through the same gas, at initial pressure p_o. The equation is plotted in Figure 8.1B. Strong shock waves are not isentropic; the entropy increases.

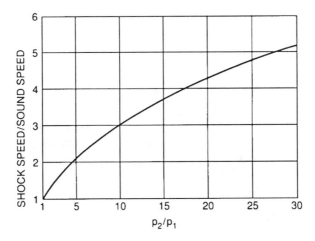

Figure 8.1B   Effect of shock strength on speed of a strong shock wave in air (Equation 8.9).

It is seen from the figure that for very small pressure differences across the shock wave, the velocity approaches that of a sound wave.

8.2 MELTING THROUGH ICE

Professor: Do we have enough facts to solve the problem?

First Student: I recall hearing that ice skating is possible because the pressure that the blades exert on ice initially at zero degrees C causes the ice to melt beneath the blades, providing a thin lubricating film. This principle must be involved in the current problem. However, we do not know the force applied to the wire, so we cannot calculate the pressure under the wire and therefore cannot solve the problem.

Professor: We were asked only to *estimate* the solution, not provide an exact answer. Furthermore, even if the pressure under the wire is as high as 100 atmospheres, this would cause the melting point of the ice to decrease by less than 1 degree, according to the data supplied. The electrically heated wire will be many degrees hotter than the melting temperature of the ice, so the temperature gradient between the wire and the melting ice surface will not be changed appreciably by this effect.

Strangely enough, however, the rate of movement of the wire through the ice can be estimated without calculating the temperature difference or the heat transfer process between the wire and the ice. Does anyone know how?

Second Student: Let's assume that as the wire moves through the ice, it melts out a channel the same width as the wire. Then the following heat balance must hold:

$$Wx = Lvxd\rho \qquad\qquad 8.10$$

where W is the power dissipation in the wire (watts per meter), x is the length of the wire in contact with ice (meters), L is the latent heat of the fusion of ice (joules per gram), v is the velocity of the wire (meters per second), d is the diameter of the wire (meters), and ρ is the density of the ice (grams per cubic meter). The length of the wire, x, is seen to cancel. Solving for v, we obtain

$$v = W/(Ld\rho) \qquad\qquad 8.11$$

Inserting the values W = 1,000 watts per meter, L = 333 joules per gram, d = 0.001 meters, and ρ = 930,000 grams per cubic meter, we obtain the result that v = 0.0032 meters per second, or 0.32 centimeters per second. Since the ice is 50 centimeters thick, it will require 50/0.32 = 156 seconds for the wire to melt through the ice.

Professor: Does anyone have any criticism of this solution?

Third Student: The solution would be correct if *all* the heat released from the wire flowed straight down into the ice. Actually, some of it must flow sideways, melting a channel somewhat wider than the wire, and some of it must flow upward, heating the cold water above it, which in turn melts more ice and makes the channel even wider.

The heat must travel from the wire primarily by conduction. But we are not sure whether the wire is in contact with liquid water or with water vapor. Furthermore, the thickness of the film of water (or water and steam) under the wire will depend on the force acting on the wire and the viscosity of water (and steam). The two-dimensional conductivity calculation will obviously be quite complex.

However, I can make an estimate of the width of the melted channel without making any detailed calculations. Refer to Figure 8.2A. If the heat from the wire flowed equally in all four directions, one-fourth of the heat

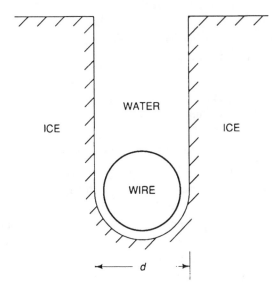

Figure 8.2A   Hot wire melting through ice.

would go to melt the ice directly under the wire and the other three-fourths would go to melting out a wider channel, so the channel width would be four times the wire width. However, since the bottom of the wire is much closer to the ice than the sides or the top of the wire, much more than one-fourth of the heat would flow downward. I say this because ice has a much higher thermal conductivity than water or steam. I shall estimate that two-thirds of the heat flows downward, and consequently the melted channel in the ice is three-halves as wide as the wire.

According to this estimate, the wire will move through the ice two-thirds as fast as previously calculated. The time to move through the half-meter distance would then be 234 seconds instead of 156 seconds.

Professor: Very good. I agree that a more precise calculation, to take into account the flow of heat in all directions, would be very difficult, but your estimate seems reasonable.

8.3 COMBUSTION WAVE IN A THERMITE MIXTURE

Professor: Is the propagation rate a linear function of the thermal conductivity of the mixture?

First Student: It seems evident to me that the heat generated in the reaction zone must be conducted through the mixture to preheat the adjacent portion to a high enough temperature so that reaction can begin in the adjacent material. This must be how the propagation occurs. I conclude that the spread rate must be directly proportional to the thermal conductivity of the mixture.

Professor: Sorry, but you are wrong. The spread rate turns out to be proportional to the square root of the thermal conductivity. I will explain this formula later, but for the moment I want to know what other variables you think are relevant in determining the propagation rate.

Second Student: Presumably the reaction between the two ingredients begins when the local temperature reaches a high enough value, which we may call the ignition temperature. Since both ingredients are originally solids, it would not seem possible for rapid reaction to occur until one of the ingredients melted. Clearly the value of the ignition temperature is an important parameter. The final temperature, which we were told is 2400 degrees C, is probably also important.

Professor: Quite so. The melting point of aluminum is 660 degrees C and that of iron oxide (magnetite) is 1,594 degrees C. We would therefore expect the ignition temperature to be at least 660 degrees C. Measurements have shown that it is about 1,100 degrees C. Any other variables?

Second Student: The mathematical formulation of the problem must involve an energy balance between the hot reacting zone and the cooler material in front of this zone. I expect that the specific heat and the density of the original mixture must be involved in this balance. In fact, I would predict that the propagation rate would be inversely proportional to the specific heat per unit volume of the material since the larger the specific heat, the longer it will take to preheat material to the ignition temperature.

Professor: Exactly correct. But we are still missing a very important variable. Because of this variable, the square-root dependence of spread rate on thermal conductivity arises.

The missing variable is the rate at which heat is liberated by the chemical reaction, once the ignition temperature is achieved. Consider a preheat zone to the left of the ignition plane (where T = 1,100 degrees C) and a reaction zone to the right of the plane (in which T rises from 1,100 to 2,400 degrees C). The more rapid the reaction, the smaller will be the reaction zone and the steeper the temperature gradient just to the right of the ignition plane.

The temperature gradient just to the left of the ignition plane will be determined by the thermal properties of the original material and by the spread rate, as we will demonstrate mathematically in a few minutes.

The crucial point is that in order for a steady-state propagation to occur, the temperature gradients just to the left and just to the right of the ignition plane must be the same (assuming no discontinuous change in thermal conductivity). Otherwise there would be an accumulation or depletion of heat at the ignition plane, which is inconsistent with a steady state.

To satisfy this condition, the propagation speed adjusts itself to a value such that the temperature gradient is continuous across the ignition plane. It can easily do so because an increase in propagation speed would cause a steepening of the temperature gradient in the preheat zone and a lessening of the temperature gradient in the reaction zone (because if the reaction requires a certain time to complete and the velocity through the reaction zone is higher, the zone must become stretched out and the temperature gradient less steep).

We see, then, that the rate of heat generation as well as the rate of heat conduction play critical roles in determining the propagation rate, and we must quantitatively analyze the interplay of these factors before predicting how the propagation rate depends on thermal conductivity.

Let's make such an analysis. Imagine a coordinate system fixed to the ignition plane. In this coordinate system, the unreacted material, at initial temperature T_0, will be advancing toward the reaction plane with the constant velocity v. (See Figure 8.3A.)

When the temperature rises to the ignition temperature T_{ig}, the chemical reaction begins and is assumed to generate heat at a constant rate per unit volume, equal to Q, until the reactants are consumed and the final temperature, T_f, is reached. The thickness of this reaction zone is L. The thermal conductivity, k; the specific heat, C; and the density, ρ, are assumed to be constant throughout.

The cross-sectional area is A; it will be seen that A will cancel out. Consider an infinitesimal slice of thickness, dx, and cross-sectional area, A, anywhere in the reaction zone. The temperature will rise because of heat released by chemical reaction within this slice; and, if the temperature gradient through the slice is not linear (i.e., dT/dx is changing), the temper-

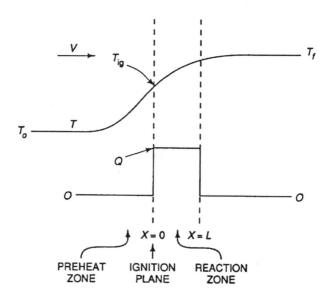

Figure 8.3A   Temperature distribututon and heat-release rate in combustion zone.

ature will also be affected by the relative rates of heat conducted into and out of the slice. Since, in the chosen coordinate system, nothing is changing with time, it follows that this energy-balance equation describes the situation:

$$v A \rho C (dT) = QA(dx) + (kA)d(dT/dx) \qquad 8.12$$

We may rearrange this equation, obtaining

$$dx = [d(dT/dx)]/[(v \rho C/k)dT/dx - (Q/k)] \qquad 8.13$$

We may integrate this differential equation, using the limits of $x = 0$ and $x = L$ for the left-hand side and the limits of $dT/dx = (dT/dx)_{ig}$ and $dT/dx = 0$ for the right-hand side. However, before doing so, we want to express L and $(dT/dx)_{ig}$ in terms of the other quantities.

From an overall balance of the process, the total rate of heat generation, QLA, must equal the difference in thermal energy of the products relative to the reactants, $v A \rho C(T_f - T_0)$. Solving for L, we have

$$L = v \rho C(T_f - T_0)/Q \qquad 8.14$$

We obtain $(dT/dx)_{ig}$ by recognizing that the heat conducted upstream across the ignition plane must be used to heat incoming material from T_o to T_{ig}. The equation expressing this process is

$$kA(dT/dx)_{ig} = v A \rho C(T_{ig} - T_0)$$

Rearranging, we obtain

$$(dT/dx)_{ig} = v \rho C(T_{ig} - T_0)/k \qquad 8.15$$

We may now integrate Equation 8.13, using Equations 8.14 and 8.15 for the appropriate limits of integration:

$$\int_0^{v \rho C(T_f - T_0)/Q} dx = \int_{v \rho C(T_i - T_o)/k}^{0} d(dT/dx)/[(V \rho C/k)(dT/dx) - (Q/k)] \qquad 8.16$$

When we integrate this equation, and insert the limits, we obtain

$$v^2\rho^2C^2(T_f - T_o)/kQ = -ln\{1 - [v^2\rho^2C^2(T_{ig} - T_o)/Qk]\} \qquad 8.17$$

Equation 8.17 gives us the desired relationship between v and the other parameters. However, this is a transcendental equation, and it is not possible to solve it explicitly for v. The relationship between v and the other quantities, as contained in Equation 8.17, is described very closely by the following empirical equation:

$$v = (1/C\rho)[\beta kQ/(T_f - T_o)]^{1/2}$$

where β is a dimensionless constant that is a function of $(T_{ig} - T_o)/(T_f - T_o)$. (When the ignition temperature is halfway between the initial and final temperatures, β is 1.6.) Let $\tau_{ig} = (T_{ig} - T_o)/(T_f - T_o)$. Then

$$\beta = 2(1 - \tau_{ig}) + 0.462(e^{1 - \tau_{ig}} - 1)/\tau_{ig} \qquad 8.19$$

Equation 8.18 shows that v is proportional to the square root of the thermal conductivity, k, and also to the square root of the heat release rate, Q. In fact, these dependences can also be deduced from a careful examination of Equation 8.17, which contains the grouping v^2/kQ in two places, whereas v, k, and Q appear nowhere else in Equation 8.17. We also see that v is proportional to $1/C\rho$, as we said earlier.

In Figure 8.3B the exact solution to Equation 8.17 is plotted (solid line) as well as the empirical approximate solution given by Equations 8.18 and 8.19 (dashed line).

You may recall that the problem also included the calculation of how much energy would have to be supplied by an ignition source to initiate a combustion wave such as this. How might we calculate this figure?

First Student: We now have a formula, Equation 8.14, by which we can calculate the thickness, L, of the steadily propagating reaction zone. We can easily compute the energy per unit area, E_{ig}, we would have to add to a slab of the material of this thickness to raise it to the ignition temperature:

$$E_{ig} = L\rho C(T_{ig} - T_o) \qquad 8.20$$

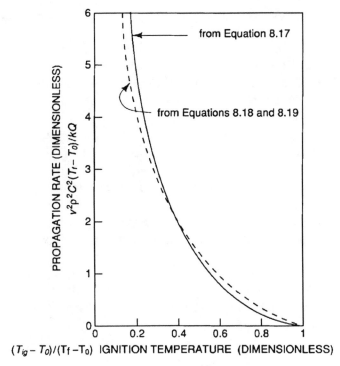

Figure 8.3B   Propagation rate versus ignition temperature.

If we substitute Equation 8.14 into equation 8.20, we obtain:

$$E_{ig} = v\rho^2 C^2 (T_f - T_0)(T_{ig} - T_0)/Q \qquad 8.21$$

I think that this equation should give a reasonable approximation to the ignition energy requirement.

Professor: I agree that Equation 8.21 gives a reasonable approximation. However, I am sure you realize that it is not exactly right because your derivation does not supply any energy to the preheat zone, and possibly supplies more energy than is needed for the reaction zone (since if you heated three-fourths of the reaction zone to the ignition temperature, e.g., it is conceivable that the released chemical energy would be enough to bring the rest of the reaction zone to the ignition temperature).

It happens that there is another way to calculate this energy requirement that is more rigorous and at the same time gives a much simpler result than Equation 8.21. Consider Figure 8.3C.

The steady-state combustion wave is converting chemical energy to thermal energy. If we forget about radiative heat loss, energy is conserved, and the sum of the chemical and thermal energy entering the combustion region must be the same as the sum of the thermal and chemical energy leaving the region.

However, because of the upstream conduction of heat, this relationship is not true within the region. As we see from the figure, the thermal energy (temperature) increases in the preheat zone before the chemical energy starts to decrease. The sum of these two energies at each point is represented by the heavy curve, and the area under the curve is the "excess energy" residing in the combustion zone as it propagates. Clearly the ignition process must supply this excess energy to initiate the process.

We can calculate this excess energy. Let e be the sum of the thermal and chemical energy per unit volume at each position in the combustion zone, and e_0 is the initial (and final) value of e. Then

$$E_{ig} = \int_{-\infty}^{\infty} (e - e_0)dx$$

8.22

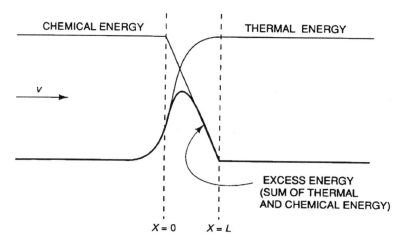

Figure 8.3C   Profile of thermal and chemical energy in combustion wave.

Consider the excess energy per unit area flowing from left to right across any plane, either in the preheat zone or the reaction zone. Its value is $v(e - e_0)$. At the same time, consider the energy per unit area being conducted from right to left across this plane, which is equal to $k(dT/dx)$. How did the material crossing the plane acquire its excess energy $(e - e_0)$? It cannot be by chemical reaction since it only changes chemical to thermal energy without changing the sum of the two. The material to the left of the imagined plane must have acquired its excess energy by conduction across the plane. This must be true regardless of where the plane is located. Accordingly,

$$v(e - e_0) = k(dT/dx) \qquad 8.23$$

Substituting Equation 8.23 into Equation 8.22 and integrating, we obtain

$$E_{ig} = \int_{T_0}^{T_f} kdT/v = k(T_f - T_0)/v \qquad 8.24$$

Equation 8.24 gives us a very simple expression for the minimum energy requirement. It is interesting to compare the prediction of Equation 8.24 with that of Equation 8.21. This comparison requires assuming a value for $(T_{ig} - T_0)/(T_f - T_0)$. If we assume a value of 0.5 for this quantity, we find that Equation 8.24 predicts an ignition energy 25 percent larger than does Equation 8.21.

The actual required ignition energy may be even larger than Equation 8.24 predicts because we have not considered the rate at which this energy would be supplied to the cold material. If it were supplied too slowly, it would be conducted too far into the material and not raise the temperature sufficiently. If it were supplied too rapidly, it might raise the surface temperature far above what is necessary and waste energy in vaporizing material.

How would you estimate the optimum rate at which to supply the ignition energy?

Second Student: I would say that the ignition energy should be supplied in a time comparable with the time it takes the steady combustion wave to move a distance L, in other words a time L/v. On the basis of Equation 8.14,

$$t = L/v = \rho C(T_f - T_0)/Q \qquad 8.25$$

If we don't know Q but we know v, we can use Equation 8.18 to get a relationship between Q and v.

Professor: That sounds reasonable, as a rough first approximation. Another way to make the estimate is to assume that the ignition energy should be applied to the surface at the same rate at which heat is conducted across the ignition plane in the steady-state propagation. That heat flux is $v\rho C(T_{ig} - T_0)$. The time would be equal to the energy needed, given by Equation 8.24, divided by the flux. The result is

$$t = k(T_f - T_0)/[v^2 \rho C(T_{ig} - T_0)] \qquad 8.26$$

The times calculated from Equations 8.25 and 8.26 are reasonably close to one another. This result is reminiscent of the reasonably close results of Equations 8.21 and 8.24, for E_{ig}. The moral seems to be that there is more than one way to skin a cat.

8.4 FLAME PROPAGATION IN A COMBUSTIBLE GAS

Professor: It must be obvious that the equations describing the propagation of a flame through a combustible gas mixture must have a lot in common with the thermite system we just discussed, in which solid reactants change into liquid products. However, there might be differences because the gas phase is involved. Who has an idea about what those differences might be?

First Student: When a flame passes through a gas, the gas expands to six or seven times its original volume. When this occurs, the expanding gas must do mechanical work against the atmosphere. This work, equal to $p\Delta V$, must be taken into account in the energy balance. Furthermore, in the equations derived in the previous section, we assumed the density to be constant. This assumption would have to be modified when dealing with a gas.

Professor: What you say is quite true. However, relatively simple modifications can be made in the equations to include these effects. Any other thoughts?

Second Student: In the thermite case, we found that the chemical rate of heat release in the reaction zone, to which we assigned a constant value Q, was quite important in influencing the propagation. We normally expect chemical reactions to increase very rapidly with increasing temperature, and the downstream side of the reaction zone may be more than 1,000 degrees hotter than the upstream side.

However, for a mixture of two materials of finite particle size, we would expect that once the ignition temperature is exceeded, the rate of diffusion of one ingredient into the other, rather than the chemical reaction rate, would control the heat release rate. The finer the particle size, the faster is the heat release rate. The rate of diffusion is much less temperature-sensitive than the rate of chemical reaction. Therefore the approximation of Q being constant would probably be much more realistic for a thermite mixture than for a gas mixture.

Professor: This argument is reasonable. To represent the rate of heat release in a gas mixture properly, we have to know which are the controlling molecular reactions in the combustion zone and how the rates of these reactions depend on temperature and partial pressures of the reactants. Such calculations can be made on large computers.

There is an effect occurring in a premixed gas flame, but not in a thermite- type flame, that complicates the calculation of partial pressures of reactants. This effect is the interdiffusion of reactants and products. Gaseous diffusion is fast enough so that combustion products, for example, can diffuse upstream to the very beginning of the preheat zone, diluting the reactants. (Condensed-phase diffusion is much slower than heat conduction, so no analogous effect occurs in thermite. Thus, Equation 8.24 in the previous section is not valid for gases.) Fortunately, by application of Fick's law of diffusion, the diffusion effect in gases can be taken into account in computer calculations.

Any other differences?

Third Student: Yes. We applied a one-dimensional model to the thermite flame, which seems reasonable. For a gas flame, it is usually not a flat surface but may be curved. We would have to allow for this curvature in our treatment. Presumably this could be done with no great difficulty, as long as the gas flame is a steady laminar flame with simple geometry.

Many practical gas flames, however, such as those burning in engines or gas- fired furnaces, are highly turbulent, and it is not clear how a steady-state one-dimensional model can be applied to such flames.

Professor: I agree. The attempt to understand simple flames must be viewed as a starting point for later research on turbulent flames, if we hope to understand them ultimately.

CHAPTER 9

A LECTURE ON DIMENSIONLESS GROUPS

In your scientific studies, you have probably encountered various dimensionless groups, with names such as the Reynolds number, the Froude number, the Mach number, the Nusselt number, etc. These groups often seem to have mysterious power to describe complex processes involving fluid motion, heat transfer, mass transfer, and chemical reaction. We're going to talk about these groups and try to lift the veil of mystery a little.

First, let's talk about fluid motion. There are mainly three kinds of forces tending to cause a fluid to move and three other kinds of forces associated with a fluid's tendency to resist moving. A number of the most common dimensionless groups are simply ratios of certain of these forces. Let's look at the forces, one at a time.

Fluid motion is often induced by a pressure acting on the fluid. Consider a differential element of fluid, of volume *dxdydz*, with a pressure difference, *dp*, acting across it in the *x* direction. The force on the element is *dpdydz*.

Fluid motion may also be induced by a body force, which is usually a gravitational force, but it might alternately be an electrostatic or electro-

142

magnetic force. If it is gravitational, its value, acting on a differential element, is $-g\rho dxdydz$. Here, g is local acceleration of gravity and ρ is density.

When there is an interface between two fluids (gas-liquid or liquid-liquid), the surface tension or interfacial tension, σ, can exert a force equal to σdx.

Now we list the forces resisting motion. The most basic is the inertial character of the fluid, which is sometimes called an inertial force. Consider a fluid element of volume $dxdydz$, which is flowing in the x direction with velocity u. If a force acts on this differential element, causing its velocity to change by an amount du, the force is calculable from Newton's second law of motion. Mass flows through the cross-section $dydz$ at the rate $u\rho dydz$. As it flows from position x to position $x + dx$, its momentum changes by $u\rho dydzdu$. Thus the force acting is $-u\rho dydzdu$. (We introduce a negative sign because du must be negative when dp is positive, and vice versa.)

Another force resisting motion is the viscous force. A *net* viscous force exists within the fluid only when the rate of change of the velocity gradient in a direction normal to the flow direction is finite. The viscous force on the element of fluid moving in the x direction, with a velocity gradient in the y direction, is given by $-\mu\partial(\partial u/\partial y)dxdz$, where μ is the coefficient of viscosity, as shown in Figure 9A.

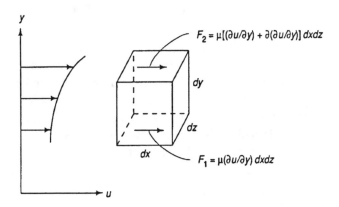

$$\text{Net force} = F_1 - F_2 = \mu\partial(\partial u/\partial y)\, dxdz$$

Figure 9A The net viscous force acting on a fluid element in a velocity gradient. (The force is the difference in the shear stress acting at the top and the bottom of the element, times the area.)

The final force we will consider is the compressive force, which is relevant only for flows in which a change of density occurs. The pressure difference, dp, needed to cause a change in density $d\rho$ is equal to $s^2 d\rho$, assuming adiabatic reversible compression, as derived in Section 8.1 (Equation 8.3). Here, s is the velocity of sound in the fluid. The force (pressure × area) is $s^2 d\rho dydz$.

Now let's see how various ratios of these forces are equivalent to various well-known dimensionless numbers. First, let's discuss the number of forces involved in any given case. If only two forces are involved, one causing motion and one resisting motion, we simply equate them. (Examples will be given later.) However, in many cases there will be more than two forces involved.

Let's discuss those cases in which three forces are involved in an important way. To be general, we denote them as F_1, F_2, and F_3. These forces, acting on any fluid element, must balance one another if other forces are negligible. Thus, we can write an equation balancing the force or forces causing motion and the force or forces resisting motion:

$$F_1 = F_2 + F_3 \text{ (or } F_1 + F_2 = F_3)$$

If we divide both sides by F_2, we obtain

$$F_1/F_2 = 1 + F_3/F_2 \text{ (or } F_1/F_2 + 1 = F_3/F_2)$$

This equation tells us that in any case in which only three forces are important, if we are able to maintain the dimensionless ratio of any two of these forces constant, the ratio of any other pair of the three forces will also automatically be constant. Thus, we can characterize the flow with a single dimensionless number. Of course, if four forces are important, we would need two dimensionless numbers to characterize the flow.

Now we discuss various dimensionless numbers.

9.1 MACH NUMBER

The Mach number, M, equal to u/s, is the ratio of the fluid velocity to the velocity of sound in the fluid. We see from the preceding relations that the ratio of the inertial force to the compressive force is $-u\rho dydzdu/s^2 d\rho dydz$. But for steady flow in the x direction with constant cross section, continuity requires that $d(u\rho) = 0$ or $ud\rho = -\rho du$. Substituting this formula in the numerator of the ratio, and canceling dy, dz, and $d\rho$, we obtain u^2/s^2 for the ratio. Thus, the square of the Mach number is equal to the ratio of the inertial force to the compressive force. If this ratio is small (small Mach number), the fluid is not compressed much while flowing and may be approximated by an incompressible fluid. If the inertial force is larger than the compressive force, the flow is supersonic and shock waves may form.

9.2 FROUDE NUMBER

The Froude number, Fr, is defined as u^2/gL. It is useful in modeling the wave resistance associated with the motion of a ship. L is a characteristic length, such as the length of the ship. If a large ship of length L_1 and a small model of length L_2 are moving through water at velocities V_1 and V_2 respectively, the wave patterns will be geometrically similar if the Froude number is the same in the two cases, i.e., $(V_1/V_2)^2 = L_1/L_2$. This formula works because when the Froude number is the same, the ratio of the inertial force associated with the liquid motion to the gravitational force controlling the height of waves is the same.

Let's demonstrate this statement. From the foregoing relations, the ratio of the inertial to the gravitational force on a differential fluid element is $u\rho dydzdu/g\rho dxdydz$. After cancellations, it becomes udu/gdx. If we are comparing a large and a small ship that are geometrically similar (one being a model of the other), it is reasonable to expect that u and du are proportional to V in the same way for the large and small ships and dx is proportional to L in the same way for the large and small ships. Accordingly, the numerator of the ratio is proportional to V^2, and the denominator is proportional to gL. Thus, the ratio is proportional to V^2/gL, which is the Froude number.

It should be noted that a pressure force is involved in this case as well as an inertial force and a gravitational force, but as previously shown, once the ratio of the inertial to the gravitational force is established, the ratio of the pressure force to the inertial force (or to the gravitational force) is automatically fixed because the three forces must be in balance.

If we want to know how the drag force, F, on the ship due to wave motion changes when we change the velocity from V_1 to V_2 and/or the length from L_1 to L_2, we write the ratio of the pressure force to the inertial force, which is $dp/(u\rho du)$. We set the drag force, F, proportional to dp and the velocity terms u and du proportional to V. Then the ratio of pressure force to inertial force is proportional to $F/\rho V^2$. For a given value of Froude number V^2/gL, the ratio of pressure force to gravitational force must be proportional to $F/gL\rho$.

Accordingly, we see how F changes for a given change in L or V (or ρ) while holding the Froude number constant.

9.3 REYNOLDS NUMBER

The Reynolds number is defined as $LV\rho/\mu$, where L is some convenient dimension associated with the flow, such as the width of a channel, the downstream distance from the leading edge of a flat plate, or the diameter of an object immersed in the flow, and V is a reference velocity at some chosen point in the flow, such as far upstream. The Reynolds number is very useful in defining "similar" flows; e.g., the wake behind an obstacle in a flow will be geometrically similar in two cases in which L, V, ρ, and μ are all different from one another, as long as the Reynolds number is the same. (See Figure 9B.) If we write the ratio of the inertial force to the viscous force from the foregoing expressions, we obtain $u\rho dydzdu/[\mu\partial(\partial u/\partial y)dxdz]$. We'll show that this ratio is proportional to the Reynolds number.

We compare two flow situations in which the boundaries are geometrically similar but of different sizes. The densities, velocities, and viscos-

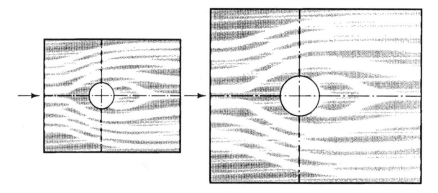

Figure 9B Flow streamlines around two cylinders of different diameters but the same Reynolds number of 25. This is an example of geometric similarity.

ities may also be different. However gravitational effects, compressibility effects, and surface tension effects are assumed to be unimportant. The flow is determined by applied pressure, inertia, and viscosity. We won't worry about pressure for the moment, but we will require the ratio of inertial to viscous force to be the same at each homologous point in the two situations.

We can assume that for the two situations, there is the same proportionality between V and u and also between V and du and between V and $\partial(\partial u)$.

We also can assume that dx, as well as dy and dz, is proportional to L for the two situations.

When we insert these various proportionalities into the ratio, some things cancel, and the original ratio is seen to be proportional to $LV\rho/\mu$, the Reynolds number.

What about the pressure force, which is involved as well as the inertial and viscous forces? Clearly, if the Reynolds number is adjusted to be the same for the two sizes L_1 and L_2, then $V_1/V_2 = (\mu_1 L_2 \rho_2)/(\mu_2 L_1 \rho_1)$. The viscosities and densities for the two sizes can be the same or different, as we wish, but once they are chosen, the velocity ratio must be as the expression says. We accomplish this requirement by applying appropriate pressure to the flow in each case to give the desired velocities. Once we have done so, we'll have similar flows.

We can determine what this applied pressure must be by calculating the ratio of the pressure force to the inertial force, which as we have seen, is $dp/(u\rho du)$. If we take dp to be proportional to Δp (the overall pressure change) and u and du to be proportional to V, the ratio in question is proportional to $\Delta p/\rho V^2$. This ratio must be invariant as long as the Reynolds number is invariant. Accordingly, $\Delta p_1/\Delta p_2 = \rho_1 V_1^2/\rho_2 V_2^2$.

9.4 GRASHOF NUMBER

The Grashof number is useful in treating flows that are driven by buoyancy. For instance, hot gas is rising, being replaced by cooler gas, but there is no forced flow; i.e., the only force applied is that of the gravitational field acting on the density difference in various parts of the flow field. The Grashof number may be defined as $gL^3\rho\Delta\rho/\mu^2$. Here, L is a characteristic dimension and $\Delta\rho$ is the density difference. (For perfect gases, it is convenient to substitute $\rho \Delta T/T$ for $\Delta \rho$, where T is the absolute temperature.)

For two buoyant-flow situations to be similar, it is necessary for the ratio of inertial to viscous forces (the Reynolds number) to be the same at each point, and also the ratio of inertial to buoyancy forces (the Froude number) must be the same. We have already seen that if one of these ratios is the same, the other must also be the same, as long as only three forces are involved. Unfortunately, each of these ratios contains the velocity, which is unknown and uncontrollable in a free buoyant flow.

We note that if we take the ratio of the Reynolds number squared to the Froude number, the velocity is canceled. Obviously, if both the Reynolds number, Re, and the Froude number, Fr, are invariant for the two situations, then the combination $(Re)^2/Fr$ must also be invariant. This ratio is the Grashof number.

However, if we use the Froude number as previously defined, we would obtain $gL^3\rho^2/\mu^2$ for the Grashof number, and comparison with the earlier-defined Grashof number shows a ρ^2 instead of $\rho\Delta\rho$. This result occurs because the Froude number was developed for treating an interface such as air-water, where the density of air is negligible compared with the density of water. When dealing with a situation involving two fluids with comparable densities (warmer and cooler gas), the equations must be modified, with $\Delta\rho$ replacing one of the ρ's.

9.5 EULER NUMBER AND BERNOULLI EQUATION

If we take the ratio of the pressure force to the inertial force, we obtain $-dp/(\rho u \, du)$. Assume that the density is constant, that we have steady flow along a streamline, and that no other forces are involved. Then, equating the two forces, we get $dp = -\rho u \, du$, which we may integrate to obtain the famous Bernoulli equation:

$$p + \rho u^2/2 = \text{constant}$$

If the two forces are not equal, the dimensionless ratio $\Delta p/(\rho u^2)$, called the Euler number, is sometimes useful.

9.6 BAROMETRIC FORMULA

If we equate the pressure force to the gravitational force, we obtain $dp = -g\rho dx$. If we assume that $\rho = Cp$ (isothermal perfect gas), where C is M/RT, a constant, then we may integrate the equation to obtain $ln\,(p/p_0) = -gC(x - x_o)$. This is the familiar barometric formula showing how atmospheric pressure would decrease with increasing altitude if temperature were constant.

9.7 WEBER NUMBER

The Weber number, defined as $L\rho V^2/\sigma$, is useful to treat the shattering of liquid surfaces into drops, the breakup of drops, and liquid motion in capillaries. If we take the ratio of the inertial force to the surface tension force, we obtain $u\rho dydzdu/\sigma dx$. If we compare two geometrically similar situations with lengths L_1 and L_2 and velocities V_1 and V_2, and take dx, dy, and dz proportional to L, and u and du proportional to V, the ratio of forces is seen at once to be proportional to the Weber number as defined.

9.8 OTHER DIMENSIONLESS NUMBERS

Enough about force ratios. There are other phenomena and other kinds of ratios that should at least be mentioned before leaving this discussion of dimensionless groups. These may involve length ratios, time ratios, physical property ratios, etc.

Let's discuss the *Prandtl number* $C\mu/k$. Here, C is the heat capacity at constant pressure. This number is important in heat transfer between a solid and a fluid, where the rate of transfer is proportional to k, the thermal conductivity of the fluid adjacent to the solid, and the resistance to heat transfer depends on the thickness of the boundary layer, which depends on the viscosity μ.

There are actually two boundary layers, the thermal boundary layer and the velocity boundary layer. It turns out that if the Prandtl number equals unity, these two boundary layers coincide, and this is the case for some fluids. On the other hand, some fluids have very high thermal conductivity relative to their viscosity (a Prandtl number much less than unity), such as liquid mercury or liquid sodium), so the thermal boundary layer extends much further into the bulk fluid than does the velocity boundary layer.

Again, the reverse is true for very viscous fluids such as heavy oils (a Prandtl number much greater than unity), the velocity boundary layer being much thicker than the thermal boundary layer in such cases. For the case of laminar flow of a cold fluid over a heated flat plate, the following equation can be derived:

$$\delta_{vel}/\delta_{th} \approx (Pr)^{\frac{1}{3}}$$

where the left-hand side is the ratio of the velocity boundary layer thickness to the thermal boundary layer thickness, and Pr is the Prandtl number of the fluid.

To see how this formula relates to heat transfer, consider steady unidirectional heat transfer by conduction, which is described by Fourier's law:

$$\text{Thermal power transferred (watts)} = kAdT/dx \qquad 9.1$$

where A is the cross-sectional area normal to the x direction and T is temperature. By introducing C, the heat capacity, and ρ, the density, this equation can be written as

$$\text{Thermal power transferred} = (k/C\rho)Ad(C\rho T)/dx \qquad 9.1a$$

where $(C\rho T)$ is the thermal content per unit volume of fluid.

Let's compare Equation 9.1a with the Hagen-Poiseuille law defining viscosity:

$$\text{Force} = \mu Adv/dx \qquad 9.2$$

But by Newton's second law, force in this case is equal to the rate of transfer of momentum across area A:

$$\text{Momentum transferred} = (\mu/\rho)Ad(\rho v)/dx \qquad 9.2a$$

Here, the right-hand side has been multiplied and divided by the density, ρ; the quantity being differentiated, ρv, is the momentum per unit volume of fluid.

When we compare Equation 9.1a with Equation 9.2a, we see that each of them gives the rate of diffusion of a quantity (heat or momentum) across a plane in terms of the gradient of that quantity. The coefficients are $k/C\rho$ and μ/ρ, respectively. If these coeffficients happen to be equal to one another, it is the same as $C\mu/k$ (the Prandtl number) being equal to unity. In this case Equations 9.1a and 9.2a would be exactly similar.

Thus, we see that the Prandtl number is the property of a fluid that is a measure of its relative ability to transfer heat and momentum across a gradient.

It might occur to you to ask about the likelihood of the Prandtl number being near unity for a fluid selected at random. It just so happens that if the fluid is a perfect gas, at any temperature and pressure and any molecular weight, the Prandtl number turns out to be a constant near unity.

The reason for this unexpected behavior emerges from the kinetic theory of gases. If V_m is the mean velocity of the molecules in the gas and L is the mean free path of these molecules between collisions, it can be deduced, as was first done by Maxwell, that the viscosity coefficient, which is a measure of the rate of transfer of momentum across a plane as driven by a velocity gradient, and the thermal conductivity coefficient, which is a measure of the rate of heat transfer across a plane as driven by a temperature gradient, are each dependent on V_m and L:

$$\mu \approx \rho V_m L/3$$

$$k \approx \rho C V_m L/3$$

These equations are so similar because the motions of the molecules, as described by V_m and L, are responsible for the transport of both momentum and heat. If we take the ratio of these two equations, we obtain $\mu C/k = Pr \approx 1$. (This relationship is only approximately true. Pr is about 0.7 for air, 0.75 for carbon dioxide, and 0.86 for ammonia.)

To show a use of the Prandtl number, let's discuss the dimensionless *Nusselt number*, which is defined as hD/k. This number is used to characterize the rate of heat exchange between a solid object and a flowing fluid at a different temperature. D is a characteristic dimension of the solid object, and h is the heat transfer coefficient in Newton's law of cooling:

$$dq/dt = hA\Delta T \qquad\qquad 9.3$$

If we visualize a thermal boundary layer of average thickness δ_{th} between the solid surface and the bulk of the fluid and we recall the Fourier law (Equation 9.1) for conductive heat transfer through this boundary layer:

$$dq/dt = kAdT/dx \approx kA\Delta T/\delta_{th}$$

we see at once when comparing this equation with Equation 9.3 that

$$h/k \approx 1/\delta_{th} \qquad\qquad 9.4$$

If we multiply both sides of Equation 9.4 by the characteristic dimension D, we obtain

$$hD/k = Nu \approx D/\delta_{th} \qquad\qquad 9.5$$

Equation 9.5 tells us that the Nusselt number is simply the ratio of the characteristic dimension of the object to the thickness of the thermal boundary layer surrounding the object.

As we have seen, the thermal boundary layer thickness is related to the velocity boundary layer thickness δ_{vel}, but these two thicknesses are not equal unless $Pr = 1$. The velocity boundary layer thickness, as a fraction of the characteristic dimension D, depends on the Reynolds number, Re, as long as the primary force acting on the fluid is a pressure gradient (not gravitation). This relationship may be written as

$$D/\delta_{vel} = f(Re) \qquad\qquad 9.6$$

where $f(Re)$ denotes some function of the Reynolds number. Also, as we have previously discussed,

$$\delta_{vel}/\delta_{th} = g(Pr) \qquad\qquad 9.7$$

where $g(Pr)$ denotes some function of the Prandtl number.

If we multiply Equations 9.6 and 9.7 together, to cancel δ_{vel}, and combine the result with Equation 9.5, we get

$$Nu = f(Re)g(Pr) \qquad\qquad 9.8$$

We may use this equation to calculate the heat transfer coefficient, h, for a given geometrical arrangement, once we have done experiments to determine the functions f and g. Equation 9.8 is the most widely used relationship in convective heat transfer.

The power of Equation 9.8 may be appreciated when we realize that the heat transfer coeficient, h, depends on seven independent variables:

- geometrical shape
- size
- fluid velocity
- fluid thermal conductivity coefficient
- fluid heat capacity
- fluid viscosity coefficient
- fluid density.

Without the aid of Equation 9.8, the correlation of heat transfer data would be very difficult.

So far we have talked about dimensionless numbers based on force ratios, physical property ratios, and length ratios. As a final example, let's consider a dimensionless number based on a time ratio, called the *first Damköhler number*.

Consider a fluid flowing with velocity V through a chemical reactor of length L. A chemical reaction is consuming a species i in the flow. The concentration of the species is c_i (moles per unit volume), and the rate at which the species is consumed is r_i (moles per unit time per unit volume).

The flow time is L/V, and the time required to consume the species is of the order of c_i/r_i. The ratio of the flow time to the reaction time would then be of the order of Lr_i/Vc_i, which is the first Damköhler number Da_I.

Clearly, if Da_I is very small compared to unity, very little reaction can occur, and the condition is described as "frozen flow." The first Damköhler number is used to predict optimum residence time in a catalytic reactor and also blowout velocity in a combustion chamber.

We have presented only a small sampling of the many useful dimensionless groups, but the general idea of how they are generated and why they are valid should be comprehensible now.

INDEX

Adiabatic compression, 97
Archimedes, 8
Arrhenius' law, 57, 90

Barometric formula, 149
Bernoulli equation, 148
Biot number, 72
Black body radiation, 90
Boltzmann's constant, 91
Boundary layer, 16-17, 20,
    149-152
Breguet range equation, 36

Celsius, 91
Charles, 91
Clausius, 120
Curie's law, 91

Diffusion coefficient, 17
Drag coefficient, 26, 34
Entropy, 116
Euler number, 148

Excess energy, 138-139
Faraday cage, 95
Film boiling, 54
First Damköhler number, 153
First law of thermodynamics, 111
Fourier's law, 150
Free energy change, 114
Froude number, 142, 145-146

Galileo, 83, 89
Grashof number, 147-148

Hagen-Poiseuille law, 150
Heat capacity ratio (specific heat
    ratio), 97, 117, 128
Heat transfer coefficient, 69, 151
Joule-Thomson effect, 95

Kelvin scale, 84, 91

Lift coefficient, 26, 34

Mach number, 142, 144-145
Maxwell's equations, 115
Michelson-Morley, 7

Newton's law of cooling, 27, 57
Newton's law of gravitation, 94
Newton's second law of motion, 88, 129
Nucleate boiling, 54
Nusselt number, 142, 151

Ohmic heating, 69
Ohm's law, 98

Parametric calculation, 29
Perfect gas law, 9, 29, 96, 127, 128
Perpetual motion, 101-102

Prandtl number, 149-152

Raoult's law, 119

Second law of thermodynamics, 115
Sonic velocity, 49, 144
Specific heat ratio (heat capacity ratio), 97, 117, 128
Stefan-Boltzman constant, 21

Trouton's rule, 90

Van't Hoff's equation, 90

Weber number, 149